# Natural Language Understanding and Cognitive Robotics

**Masao Yokota**
Department of Intelligent Information Systems Engineering
Fukuoka Institute of Technology, JAPAN

CRC Press is an imprint of the
Taylor & Francis Group, an **informa** business
A SCIENCE PUBLISHERS BOOK

CRC Press
Taylor & Francis Group
6000 Broken Sound Parkway NW, Suite 300
Boca Raton, FL 33487-2742

First issued in paperback 2021



Version Date: 20190802

ISBN-13: 978-0-367-36031-3 (hbk)
ISBN-13: 978-1-03-208748-1 (pbk)



Library of Congress Cataloging-in-Publication Data

Names: Yokota, Masao, 1954- author.
Title: Natural language understanding and cognitive robotics / Masao Yokota, Department of Intelligent Information Systems Engineering, Fukuoka Institude of Technology, Japan.
Description: First edition. | Boca Raton, FL : CRC Press/Taylor & Francis Group, 2019. | Includes bibliographical references and index. | Summary: "Describes the method for robots to acquire human-like traits of natural language understanding (NLU), the core of which is mental image directed semantic theory (MIDST). It is based on the hypothesis that NLU by humans is essentially processing of mental image associated with natural language expressions. MIDST is intended to model omnisensory mental image in humans and provide a knowledge representation system for integrative management of knowledge subjective to cognitive mechanisms of intelligent entities, e.g. humans and robots, with a systematic scheme for symbol-grounding. The book is aimed at researchers and students interested in artificial intelligence, robotics, or cognitive science. Based on philosophical considerations, this will also have an appeal in linguistics, psychology, ontology, geography, and cartography"-- Provided by publisher.
Identifiers: LCCN 2019029576 | ISBN 9780367360313 (hardback ; acid-free paper)
Subjects: LCSH: Natural language processing (Computer science) | Human-machine systems. | Autonomous robots.
Classification: LCC QA76.9.N38 Y65 2019 | DDC 006.3/5--dc23
LC record available at https://lccn.loc.gov/2019029576

**Visit the Taylor & Francis Web site at**
**http://www.taylorandfrancis.com**

**and the CRC Press Web site at**
**http://www.crcpress.com**

# Preface

With the remarkable progress in robotics and artificial intelligence today, it is likely that robots will behave like ordinary people (necessarily of good will) in the not-so-distant future and that a symbiotic world will come to be, where humans and robots will be interact amicably with each other. For this to happen it is important robots must be developed with an understanding of what people do (e.g., see, hear, think, speak, ...). The most essential capability for the robots to satisfy this requirement will be that of integrative multimedia understanding, which is performed very easily by people and which brings forth to them total comprehension of information pieces separately expressed in miscellaneous information media, such as language, picture, gesture, etc.

This book describes the challenges of providing robots with human-like capability of natural language understanding as the central part of integrative multimedia understanding. The core of the methodology is my original theory named mental image directed semantic theory, based on the understanding that natural language understanding by humans is essentially the processing of mental images associated with natural language expressions, i.e., mental-image based understanding. Mental image directed semantic theory is intended to model omnisensory mental image in humans and to provide knowledge of representation system in order for integrative management of knowledge subjective to cognitive mechanisms of intelligent entities, such as humans and robots, based on a mental image model and its description language named mental image description language. This language works as an interlingua among various kinds of information media and has been applied to various versions of the intelligent system named IMAGES. Among them, its latest version conversation management system, is a good example of artificial general intelligence performing integrative multimedia understanding as well. Conversation management system simulates human mental-image based understanding and comprehends users' intention through dialogue in order to find solutions, and return responses in text or animation.

Viewed from the perspective of computer science, almost all terminologies of human cognition, such as natural language understanding and mental imagery, have never been given concrete definitions. I found a lot of literature concerning mental imagery beyond my comprehension because of its abstractness, sometimes, cloud grabbing stories for me. In mental image directed semantic theory, however, I always try to introduce them formalized definitely on the standpoint of computation in association with the mental image model, as well as employing plenty of diagrams and illustrations for intuitive comprehension, which is also the case for this volume. Mental image directed semantic theory is a hypothesis, still evolving under my philosophical insight, to be presented throughout this book. I believe that any kind of philosophical consideration is essential to providing a hypothesis with a certain perspective for its potentiality, especially when the very problem is too profound to be solved at a time. The expected readers of this volume are researchers or students who dream to realize *true* natural language understanding in cognitive robots. It will be my great pleasure if this book can inspire them with some hints to solve their own problems.

I convey my special thanks to Dr. Khummongkol, Rojanee (University of Phayao, Thailand) for her help in developing an excellent conversation management system as my PhD. student in Japan.

Masao Yokota
31 December, 2018

# Contents

# 1
# Introduction

The long history of science and remarkable progress in artificial intelligence and robotics today may well predict that the day will come in the near future when artifacts will catch up with or overcome people in every aspect. However, it is considered that the prediction should be only the case for the fields where their capabilities can be acquired simply from numerous pairs of input (or stimulus) and output (or response). For example, pattern recognition by deep neural networks is one of the most successful cases. On the other hand, such human abilities as natural language understanding, where the cognitive *processes* must be clarified in advance, have not been successfully transplanted to machines yet. Yokota, the author, has been working on natural language understanding by computers with his belief that people understand an expression in natural language intrinsically by operating mental images evoked by it, and developed an original semantic theory named mental image directed semantic theory (conventionally, abbreviated as MIDST).

This book describes mental image directed semantic theory with its philosophical background and presents a methodology challenging to provide robots with human-like capability of natural language understanding centered at integrative multimedia understanding as the most essential ability that leads people and machines to totally coherent and cohesive comprehension of information pieces separately expressed in linguistic or non-linguistic information media. The methodology is applied to three robots named 'Poci', 'Robby', and 'Anna', respectively, where the first one is a real dog-shaped robot and the rest are imaginary humanoid robots. In general, robots can acquire or have semantics of human language (i.e., natural language) differently from humans due to the hardware or the software specific to them and therefore their natural language understanding capabilities can also be specific to them as detailed later. The work presented in this volume is mainly led by an account of possible real life scenario of the future, describing an old man (Taro) in the near future with physical disabilities, and his interactions with a robot assistant (Anna) intended as an ideal humanoid robot.

## 1.1 Anna—An Ideal Home Robot

Anna comes to Taro's home to live with and help him. Taro welcomes Anna with a big smile (see Fig. 1-1). Anna is a lady-shaped robot, looking so cute. Taro is a retired professor from a university where he was majoring in robotics and designed Anna technically as a home robot. At that time, Taro is old and physically handicapped, so he cannot walk by himself and is usually confined to a wheelchair. At last, confronting inconveniences in various scenes of his daily life, he is convinced that he should order a real Anna from a robot maker and the day comes when she appears in front of him at home. One day, Taro tells Anna "I want to call Tom". Then, she asks him "Where is your cell-phone?", he replies "On my bed" and she says "Certainly, I'll bring back it to you". Saying so, she begins to walk toward his bed. Finally, she comes back to him with his cell-phone, exclaiming in delight "Here, you are!"

Judging from this scene, Anna can at least understand multiple types of information media in a certain integrative way, such as natural language expressions and sensory data from the external world, and find/solve problems based on such integrated information. That is, Anna is provided with human-like cognitive abilities in the full range, so-called artificial general intelligence, surely at least with the ability of integrative multimedia understanding necessarily involving natural language understanding and problem finding and solving.

The terminology artificial general intelligence, however, has no clear definition yet, but it refers to artificial intelligence for a machine to perform human cognitive abilities in the full range, while conventional artificial intelligence can only accomplish problem solving or reasoning in specific task domains, such as the block world of SHRDLU (Winograd, 1972), the first natural language understanding system so far reported (Copeland, 2018). Anyhow, such a description of artificial general intelligence is not immediately practical for computation. More precisely, artificial general intelligence yet remains in an imaginary robot in the world of science fiction, like *Data*, the humanoid robot in the *Star Trek* television series.

**Fig. 1-1.** Anna and Taro.

## 1.2 Intuitive Human-Robot Interaction

Remarkable progress in robotics and artificial intelligence today may well predict that robots like Anna will be behaving just like ordinary people in the near future and that a symbiotic world of the both entities will come true where they will be enjoying harmonious interaction with each other. For such a desirable symbiosis, robots should be at least required to comprehend and perform well enough what ordinary people do (e.g., see, hear, think, speak, etc.). The most essential capability needed for the robots to satisfy this requirement will be that of integrative multimedia understanding. This kind of cognitive behavior is performed very easily by people and brings forth to them total comprehension of information pieces separately expressed in miscellaneous media, such as language, picture, gesture, etc., accompanied with problem finding and solving based on such integrated information.

Recently, there have been various types of real or virtual robots developed as artificial partners. However, they are to play their roles according to programmed reactions to stimuli and have not yet come to perform human higher-order mental functions, such as integrative multimedia understanding. That is, the symbiosis mentioned above poses a very high hurdle before robots at the present stage, but it could be a very convenient world for people because the robots there will be wise enough to understand and support them in every way, going as far as to guess their implicit beliefs and emotions.

It should be noted that for a robot to have human-like emotions and to be intelligent enough to learn and understand human emotions are very different things. Mr. Spock, who is not a robot but an emotionally-detached, logic-driven alien officer in Star Trek, said **"Insults are effective only where emotion is present."** (*Star Trek*, season 2, episode 2, "Who Mourns for Adonais?", 1967), but he must have a good knowledge of human emotional propensities, judging from his quote **"After a time, you may find that having is not so pleasing a thing after all as wanting. It is not logical, but it is often true."** (*Star Trek*, season 2, episode 1, "Amok Time," 1967). That is, Mr. Spock is emotionless but has a good model of human emotion as part of his *knowledge*.

The author has been working on integrative multimedia understanding in order to facilitate intuitive human-robot interaction, as shown in Fig. 1-2, that is, interaction between non-expert, ordinary people and home robots (Yokota, 2005, 2006). For ordinary people, natural language is the most important among the various communication media because it can convey the exact intention of the emitter to the receiver due to the syntax and semantics common to its users (Austin, 1962; Searle, 1969; Vieira et al., 2007). This is not necessarily the case for another media, such as gesture, and so natural language can also play the most crucial role in intuitive human-robot interaction. For example, robotic imitation, as shown in Fig. 1-3, will lead us to instant comprehension of this fact, where the robot trainer is demonstrating in front of Robby without/with verbal hint.

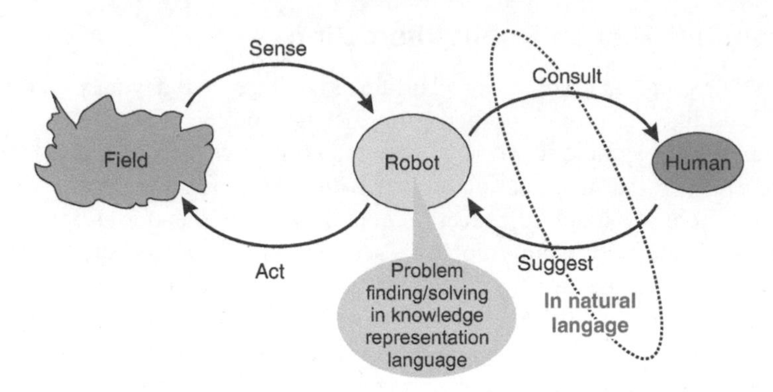


**Fig. 1-2.** Intuitive human-robot interaction based on integrated multimedia understanding.

**(a)** Without verbal hint

**(b)** With verbal hint

**Fig. 1-3.** Robotic imitation as intuitive human-robot interaction.

In order to realize integrative multimedia understanding intended here, it is essential to develop a systematically computable knowledge representation language (KRL) as well as representation-free technologies (Brooks, 1986), such as neural networks for unstructured sensory/motory data processing (i.e., *stimulus/ response processing*). This type of language is indispensable to *knowledge-based* processing such as *understanding* sensory events, *planning* appropriate actions and *knowledgeable* communication with ordinary people in natural language. Therefore, such a KRL needs to have at least a good capability of representing spatio-temporal events in good correspondence with human/robotic sensations and actions in the real world, which is possibly in the same line of 'common coding approach to perception and action' (Prinz, 1990).

Overviewing conventional methodologies, almost all of them have provided robotic systems with such quasi-natural language expressions as 'move (Velocity, Distance, Direction)', 'find (Object, Shape, Color)', etc., for human instruction or suggestion, uniquely related to computer programs for deploying sensors/motors as their semantics (e.g., Coradeschi and Saffiotti, 2003; Drumwright et al., 2006). These expression schemas, however, are too linguistic or coarse to represent and compute sensory/motory events in such an integrative way as the intuitive human-robot interaction intended here. This is also the case for artificial intelligence planning ('action planning'), which deals with the development of representation languages for planning problems and with the development of algorithms for plan construction (Wilkins and Myers, 1995).

In order to challenge a complex problem domain, the first thing to do is to design/select a certain KRL suitable for constructing a well-structured problem formulation, namely, a representation. Among conventional KRLs, the ones employable for first order logic have been the most prevailing because of good availability of deductive inference engines intrinsically prepared for computer languages such as Python. According to these schemes, for example, the semantic relation between '*x* carry *y*' and '*y* move' is often to be translated into such a representation as $(\forall x,y)\ (carry(x,y) \supset move(y))$. As easily imagined, such a *declarative* definition will enable an natural language understanding system to answer correctly to such a question as "When Jim carried the box, did it move?" but it will be of no use for a robot to recognize or produce any external event referred to by '*x* carry *y*' or '*y* move' in a dynamic and incompletely known environment unlike the Winograd's block world (Winograd, 1972). That is, this type of logical expression as is can give only combinations of dummy tokens at best. For example, $carry(x,y)$ and $move(y)$ are substitutable with $w013(x,y)$ and $w025(y)$, respectively, which do not represent any word concepts or meanings at all but are the *coded names* of such concepts or meanings. If you find any inconvenience with this kind of substitution, that is due to being without symbol grounding (Harnad, 1990) on your lexical knowledge of English. Schank's Conceptual Dependency theory (Schank, 1969) was an attempt to decrease paraphrastic variety in knowledge representation by employing a small set of coded names of concepts called conceptual primitives, although its expressive power was very limited.

The fact above destines a cognitive robot to be provided with *procedural* definitions of word meanings grounded in the external world, as well as *declarative* ones for reasoning in order both to work its sensors and actuators appropriately and to communicate by natural language with humans properly. Therefore, it is noticeable that some certain interface must be employed for translation between *declarative* and *procedural* definitions of word meanings, where the problem is how to realize such a translator systematically. Conventional KRLs, however, are not so viable of such systematization because they are not so cognitively designed, namely, not so systematically grounded in sensors or actors. **That is, they are not provided with their semantics explicitly but implicitly grounded in natural language word concepts that can be interpretable for people but have never been grounded in the world well enough for robots to cognize their environments or themselves through natural language expressions.**

Quite distinguished from the conventional approaches to natural language understanding (Weizenbaum, 1966; Winograd, 1972; Schank and Abelson, 1977; Ferrucci, 2004; Winograd, 2006), mental image directed semantic theory (e.g., Yokota, 2005) has proposed a dynamic model of human sensory cognition, yielding omnisensory image of the world and its description language named 'mental image description language ($\boldsymbol{L}_{md}$)'. This formal language is one kind of KRL employed for predicate logic.

## 1.3 Integrative Multimedia Understanding and Natural Language Understanding

Just reflecting our own psychological experiences, there are two major thinking modes in humans (Paivio, 1971). One is our inborn non-linguistic thinking which starts from infancy, especially experienced in visuospatial reasoning, and the other is our postnatal language-based thinking, acquired as we grow up, which enables us to perform higher order abstract reasoning. Non-linguistic thinking involves mental operations on purely non-linguistic internal representation (e.g., Shepard and Metzler, 1971), namely, mental image being perceived from the environment (i.e., perceptual image) or being evoked inside the brain (i.e., conceptual image), and so often on mixtures of both types (more exactly, perceptual image complemented by conceptual one), not limited to visual but omnisensory. For example, consider the picture in Fig. 1-4a, so called 'Rabbit and Duck', and its left and right parts in Fig. 1-4b and 4c. Figure 1-4a is surely ambiguous, reminding us of both Rabbit and Duck alternately, but each of its parts is not. Therefore, it is thought that this psychological phenomenon should be due to our scanning directions upon this picture and the associated conceptual images, namely, Rabbit from the left and Duck from the right. That is, we are to complement the perceptual

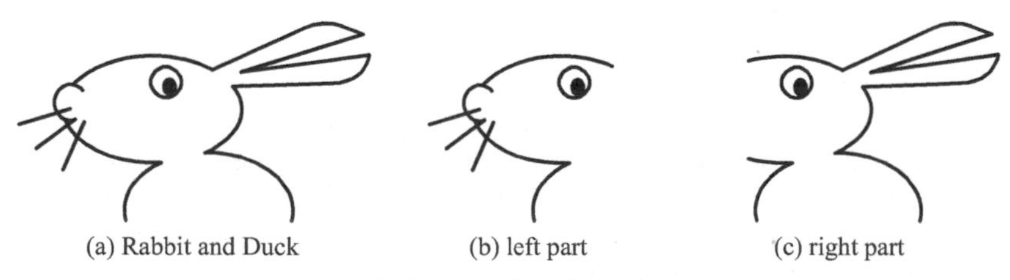

(a) Rabbit and Duck (b) left part (c) right part

**Fig. 1-4.** Rabbit and Duck, and its parts.

images at the other parts with the conceptual images induced from our knowledge in association with the perceptions. As easily understood, this predictive function sometimes leads us to misrecognition as well as efficient recognition.

On the other hand, linguistic thinking requires the ability of bidirectional translation between external representation (i.e., expression in natural language) and internal representation of ideas. In general, this ability is called natural language understanding while it is usually accompanied with reasoning, based on internal representation such as question answering. For example, try to interpret the assertion S1-1 and answer to the questions S1-2 and S1-3. Perhaps, without any exception, we cannot answer the questions correctly without reasoning based on the mental images evoked by these expressions. This kind of reasoning belongs to what are required for the Winograd Schema Challenge (Levesque, 2011) that would discourage conventional natural language understanding systems adapted for the Turing Test (Turing, 1950), even the renowned quiz champion artificial intelligence, Watson by IBM (Ferrucci et al., 2010). The Winograd Schema Challenge is a more cognitively-inspired variant of the Textual Entailment Challenge (Dagan et al., 2006) to eliminate cheap tricks intended to pass the Turing Test, which is essentially based on behaviorism in the field of psychology.

(S1-1) Mary was in the tram heading for the town. She had a bag with her.

(S1-2) Was the tram carrying Mary?

(S1-3) Was the bag heading for the town?

For ordinary people, the ability of natural language understanding is doubtlessly indispensable to comprehensible communication with others. Anyhow, what is very remarkable here is that they can deploy these two thinking modes *seamlessly* and almost unconsciously. This fact leads to a very natural conclusion that intelligent robots thinking like humans, so-called cognitive robots in this book, should be capable of understanding both linguistic and non-linguistic information media in a certain *integrative* way, which is the very thing, integrative multimedia understanding, to be pursued here.

It is considered that integrative multimedia understanding is possible only when mental (or internal) representation of the world can work both symbolically and non-symbolically. Here, such internal representation is called quasi-symbolic image representation. A quasi-symbolic image is like a pictogram or a more abstract token or scheme often tagged with linguistic identifications. Consider Fig. 1-5 where a person observes his environment and recognizes what is happening there. This is an example of human selective attention, resulting in a piece of 'direct knowledge' so called here, where only the objects of interest, namely, flower-pot, lamp, chair, cat, and box appear in the live image, but the person and the ball outside the window do not appear because they are of no interest to him.

Figure 1-5 shows that the piece of direct knowledge consists of quasi-symbolic images as dots (tagged with their names) in a 4D context and also as a

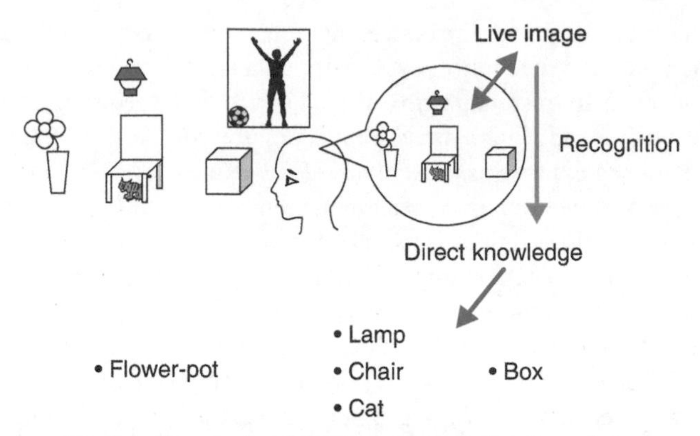


**Fig. 1-5.** Transduction from live image into direct knowledge.

quasi-symbolic image that work both as symbols and as virtual realities of actual matters. Thanks to direct knowledge as quasi-symbolic image representation, we can answer such questions as 'what is between the flower-pot and the box?' very easily without the live image.

For example, the author's mental image evoked by the expression 'X (be) between Y and Z' is as shown in Fig. 1-6, consisting of four quasi-symbolic images. They are three real objects as disks and one imaginary object as an area enclosed by a dotted line with the verbal tokens 'X', 'Y', 'Z', and 'X (be) between Y and Z' tagged, respectively, subordinated by the quasi-symbolic images and verbal tokens (i.e., 'btwn_*i*') for its instances. By applying Fig. 1-6 to Fig. 1-5, we can generate the two expressions S1-4 and S1-5, as if *seamlessly*, from the external scene depicted in Fig. 1-5. This is a good example of seamless switching between linguistic and non-linguistic thinking modes via quasi-symbolic images with verbal tokens. Such a switching mechanism is one of the functions of integrative multimedia understanding to be implemented in Anna.

(S1-4) The chair is between the flower-pot and the box.

(S1-5) The chair is between the lamp and the cat.

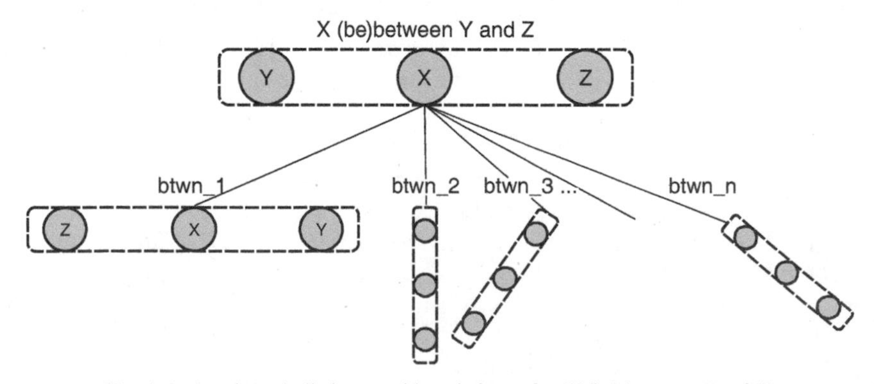


**Fig. 1-6.** Quasi-symbolic image with verbal tags for '*X* (be) between *Y* and *Z*'.

As mentioned above, direct knowledge of a perceived matter is not its pure analogue like a picture but articulated into a structure of its significant features and can offer its information in a level of abstractness appropriate enough for necessity. For example, the map of the furniture pieces in a room can be reproduced from the direct knowledge of them as a spatial arrangement of their names or labels reflecting their actual spatial relations. The author thinks that every kind of knowledge of the world is based on such direct knowledge, not limited to visual but omnisensory, and this thought is the starting point of mental image directed semantic theory. To his best knowledge, no conventional methodologies have been reported to represent and compute multimedia information in an integrative way, but linguistic information exclusively (Schubert, 2015a).

## 1.3 Knowledge and Cognition

A symbol system to formalize information about the world to compute is generally called 'knowledge representation language (i.e., KRL)' (Davis et al., 1993; Brachman, 1985). So far, there have been several kinds of knowledge representation methods or technologies proposed, and thereby a considerable number of KRLs, such as KL-ONE (Brachman and Schmolze, 1985), CycL (Guha and Lenat, 1990), and OWL (Tsarkov and Horrocks, 2006), were invented for formal ontology of the world, namely, formalization of (concerned portions of) the world with such an intention so as to facilitate inferences involved. However, almost all such ontological commitments about the world, for example, CYC project (Lenat and Guha, 1990), were quite objective, namely, without any reflection of human cognitive processes or products in spite of the fact that ordinary people live the greater part of their casual life based on their own subjective and intuitive knowledge of the world acquired through their inborn sensory systems. Actually, according to the author's long-term study, it is of no doubt that people's concepts about the world expressed in natural language are very much affected by their cognitive propensities, which is quite natural, considering that they are genuine cognitive products in humans. For example, people often sense continuous forms among separately located objects, known as *spatial gestalts* in the field of psychology, and refer to them by such an expression as 'Four buildings stand in *square*.' For another example, what should be the ontology of a *slope* to make machines understand such a person's intuitive utterance as 'The path *sinks* to the basin.' or 'The path *rises* from the basin.'? Apparently, they are affected by human active perception of the physical world. These facts also imply that knowledge representation without reflecting human cognition is not very useful for interaction between ordinary people and home robots, especially by natural language, so called here, intuitive human-robot interaction. That is, approaches to natural language understanding should be much more cognitive than behavioristic methodologies in natural language processing, such as statistical machine translation (Manning, 1999; Koehn, 2009; Navigli, 2009), recently prevailing, based on certain text corpora. In other words, designing a cognitive robot can be knowing human cognition (Wilson and Foglia, 2017), which is challenging yet.

According to author's own experiences, he inevitably employs mental images when he understands natural language descriptions about the places he will visit for the first time, which leads to his proposal of the knowledge representation language $\boldsymbol{L}_{md}$ in mental image directed semantic theory (i.e., MIDST). Quite distinctively from others, $\boldsymbol{L}_{md}$ is designed to simulate mental image phenomena subjective to humans, and he has found that the word concepts concerning space can be formalized systematically in $\boldsymbol{L}_{md}$ by introducing spatial gestalts as virtual objects and a theoretical model of human active sensing. Despite the tremendous amount of work concerned, mental image or natural language understanding, even the meaning of *meaning* itself, has not been given any concrete definition. Therefore, MIDST is still a hypothesis and this book is intended to show its potential validity. The following chapters describe MIDST and its application to natural language understanding for cognitive robotics (Levesque, 2011), namely, knowledge representation and computation to facilitate intuitive interaction between people and robots through natural language. It is noticeable that the semantics of KRLs themselves are conventionally not given in a comprehensive way. That is, conventional KRLs, except $\boldsymbol{L}_{md}$, are not provided with their semantics explicitly but implicitly grounded in natural language word concepts that can be interpretable for people but have never been grounded in the world well enough for robots to cognize their environments or themselves through natural language expressions.

What the author intends to present in this book is digested in two points concerning KRLs. One is from the viewpoint of symbol grounding. Apart from conventional approaches to natural language understanding with input and output in text, natural language semantics for cognitive robots to act properly should be grounded in the sensors and actuators specific to them. For example, we must have a robot behave in the environment to people's commands, such as 'Put the book on the chair between the table and the desk', where the robot must judge whether or not the environment will permit its planned behavior according to the implemented algorithms. In other words, certain simulation is needed in advance to its real action, based on the sensory information about the environment. If there are any obstacles, the robot must avoid or remove them in the course of its action. That is, the robot must be provided with a mechanism to assess the environment in advance to its real action partially or totally based on certain internal representations of the environment. The other is from the viewpoint of human-robot interaction. Robots are always to interpret people's utterances both in human-specific and robot-specific semantics. Otherwise, both the entities would confront a serious cognitive divide that would hinder their mutual comprehension. For example, people could order to a four-legged dog-shaped robot to 'Wave your left *hand*,' where the robot would have to interpret this command so that they are asking it to wave its left *foreleg*. For another example, people can comment on spatial arrangement of objects like 'the Andes mountains run north and south.' Even if a robot could not understand this comment in its own semantics, it would have to interpret the utterance in human-specific semantics in order to hold good conversation with them. The formal language $\boldsymbol{L}_{md}$ is designed for systematic representation and computation of knowledge in the framework of the proposed human mental image model in order to satisfy these requirements for KRLs.

# 2
# Natural Language Processing Viewed from Semantics

Natural language processing refers to all fields of research concerned with the interactions between computers and human (natural) languages. However, it is often associated specifically with programing computers to process and analyze large amounts of natural language data where natural language generation is detached from it or its accessory without much elaborated sophistication. Historically, natural language processing started with machine translation (Hutchins, 2006) which necessarily requires natural language generation. Here, from the standpoint of natural language understanding, natural language processing means especially what involves a pair of processes to analyze text into certain intermediate representation and to generate text from it (directly or indirectly), where the languages for analysis and generation can be either the same or not. Automatic paraphrasing (e.g., Schank and Abelson, 1977) and text summarization within one language (e.g., Yatsko et al., 2010), and Kana/Kanji (Japanese syllables/ Chinese characters) conversion in Japanese (www.pfu.fujitsu.com/hhkeyboard/ kb_collection/#nec) are examples of the former. On the other hand, machine translation is the typical example of the latter where cross-language operation is inevitably performed by computers. This chapter considers machine translation as the representative natural language processing and emphasizes the necessity of natural language understanding as the genuine natural language processing.

## 2.1 Trends in Machine Translation

From the 1950s, when the earliest machine translation systems emerged, to the 1980s, there were two types of approaches to cross-language operation, namely, interlingual and transfer-based approaches. In the former, the source language is translated into a certain interlingua, and the target language is generated from the interlingua independently of the source language. In the latter, each pair of source language and target language requires a module, called transfer component, to exchange their corresponding intermediate representations, for example, language-specific dependency structures as grammatical descriptions. Both the techniques are referred to as rule-based machine translation but the transfer approach has been prevailing because, from the technical viewpoint, it is very difficult to develop an interlingua except in limited task domains where expressions are well formulated or controlled. The most serious problem of rule-based machine translation is that it needs extensive lexicons with morphological, syntactic, and semantic information, and large sets of rules, almost all of which are inevitably hand-written.

In the late 1980s, natural language processing based on statistics, so-called statistical natural language processing, became another major trend. In principle, statistical natural language processing is centered on machine learning and driven by statistical inferences automatically acquired from text corpora instead of hand-written rules. Therefore, its accuracy depends on the learning algorithms involved and the quality of the corpora. In particular, statistical machine translation requires multilingual corpora as collections of translations of *high quality*. Here, what is meant vaguely by the phrase 'high quality' is always a serious problem for machine learning. For example, what is learnt by free translations must be quite different from what is learned by literal translations. So, which is higher in quality? The less vague interpretation of the phrase is that one corpus is higher in quality than another if it makes machine learning more *successful* in machine translation (or natural language processing) than the other, where, in turn, a certain authorized metric is required in order to evaluate the successfulness in machine translation, including automated means, such as BLEU, NIST, METEOR, and LEPOR (e.g., Han et al., 2012).

So far, a considerable number of automatic translation (or machine translation) systems, either operational or experimental, have been developed. It can be very natural to consider that machine translation should require the problems of natural language understanding to be solved first, considering that people decode the meaning of the source language text and re-encode this meaning in the target language during their translation process. However, the heart of the matter is also the knowledge representation language to represent meaning, which is traditionally called 'interlingua' in the field of machine translation. As already mentioned, it is very difficult to design such a knowledge representation language, especially so as to be perfectly language-free. Therefore, the interlingual machine translation approach has been made operational only in relatively well-defined domains, such as device manuals (Nyberg and Mitamura, 1992), and instead the transfer-based machine translation approach has been made more operational, where meaning representation depends partially on the language pair involved in the translation. These two traditional methodologies belong to the category of rule-based approaches controlled by certain linguistic knowledge implemented as rules. Currently, other approaches without such rules but based on big data, called multilingual corpora, are prevailing, for example, statistical machine translation (Koehn, 2009) and neural machine translation (Bahdanau et al., 2015) although they are going farther from natural language understanding than the two traditional approaches. However, it is noticeable that the quality of machine translation can be indifferent of natural language understanding, for instance, example-based translation (Nagao, 1984), on the assumption that a sufficient number of good examples of translations to simulate are available in advance.

We have developed an machine translation system named ISOBAR (Yokota et al., 1984a) which translates Japanese weather reports into English. In the domain of meteorological reports, TAUM-METEO is the best-known operational system (Chandioux, 1976), performing English-French translation, which is classified as one of the transfer systems. The translation by ISOBAR is experimental, but its most remarkable feature is that target language sentences are generated only from interlingual meaning representation in $\boldsymbol{L}_{md}$ of input source language sentences. That is, ISOBAR has no transfer component. In this approach, one sentence can be translated into multiple target language sentences which are in paraphrase relation. During semantic processing of source language sentences, ISOBAR is designed to reject such sentences as are semantically anomalous or contradictory to its knowledge specific to weather reports, unlike Wilks' model (Wilks, 1972), so-called Preference Semantics based on a set of semantic primitives (probably intended for general purpose). Preference Semantics calculates the semantic plausibility of a sentence based on the number of internal preferences of its parts (normally verbs or adjectives) satisfied and determines its most coherent interpretation. The difference between Wilks' and ISOBAR in this point is greatly due to whether or not the application domains are strictly limited.

Anyway, we cannot rely so heavily on the results of machine translation or natural language processing bypassing semantic evaluation based on natural language understanding. The most serious problem of machine translation systems without text understanding is that we cannot tell when they give us mistranslations because we need them especially when we are unfamiliar with the target language.

## 2.2 Case Study of Current Machine Translation Systems (as of October, 2018)

The author has a very impressive memory of machine translation. In June 2017, he visited Italy and stayed at a B & B in a rural district where it was quite rare for him to encounter English speakers, to say nothing of Japanese speakers. Knowing that he did not understand Italian, the landlady tried to communicate with him in English by using an machine translation utility through her smart phone. She showed him the translations which he often found to be inaccurate. Anyway, he very much enjoyed imagining what she intended to say through them, but he wondered what would have happened to him if the machine translation system had translated her intention but making sense in a wrong way, e.g., 'It is fine to go out with any equipment' instead of 'Fine for taking out any equipment'.

Now, suppose that you are traveling by yourself in a foreign country where the conventional language is quite unfamiliar to you. In preparation, you are carrying a machine translation utility with you which is announced to be available for bidirectional translation between the foreign language and yours. Then, what will you do to know thc precision of its translation?

If the author were you, he would make it translate its output again back into his mother language each time, although the quality of the forward translation can be distorted in the backward. Such an examination is conventionally known as round-trip translation (or back-and-forth translation) and is very quick and easy, especially for laypeople in certain practical use scenarios. However, note that this methodology is inappropriate for any serious study of the quality of a single machine translation system because it inevitably involves two systems, namely, onc for source language to target language and the other for target language to source language. From another viewpoint, this kind of trial can be useful to consider how source language analysis and target language synthesis should be elaborated for indirect translation (e.g., Japanese-English-Thai) based on the sentence-transfer method.

Out of such a scientific curiosity of the author's, round trips were tried on several machine translation systems currently for public use on the web in order to know how they can preserve the *meaning* of the original source language input or how they can cope with semantic problems involved such as semantic ambiguity.

Three systems, Google Translate, Systran and Excite, were investigated with English as the source language and Japanese as the target language. As well known, the first system is a statistical or neural machine translation system on the basis of statistical models automatically acquired from bilingual text corpora and the second one is rule-based, almost hand-crafted and partially automated. The last one, in the author's experience as a user, is also rule-based, probably hand-crafted. It is easily imaginable that translation between similar languages could be rather successful without natural language understanding and therefore the pair of English and Japanese was selected, being greatly different from each other. One of serious problems involved in machine translation is how to dissolve ambiguities in source language texts. Therefore, a closer look was taken, especially on how they cope with this problem in processing the sample sentences I-1–I-8 as follows.

(I-1) With the box, the boy follows the girl.

(I-2) The boy with the box follows the girl.

(I-3) The boy follows the girl with the box.

(I-4) Bring me the candy on the desk which is very sweet.

(I-5) Look at the car on the road which is running very fast.

(I-6) Tom was with the book in the bus running from Town to University.

(I-7) Tom was with the book in the car driven from Town to University by Mary.

(I-8) Tom kept the book in a box before he drove the car from Town to University with the box.

These sentences are arranged to be ambiguous and their linguistic features are summarized as follows. Firstly, I-1, I-2, and I-3 are simple sentences involving the prepositional phrase 'with the box' placed differently each time. As easily understood, I-1 is not ambiguous either syntactically or semantically. On the other hand, I-3 is ambiguous both syntactically and semantically while I-2 is ambiguous syntactically but not semantically. Secondly, I-4 and I-5 are complex sentences, each of which is a simple sentence followed by a simple relative clause. Each of them involves syntactic ambiguity which is dissolvable if the system is designed to exploit the ontological information from the relative clause. Lastly, I-6, I-7, and I-8 are rather complicated sentences with multiple prepositional phrases, containing subordinate clauses introduced by the present particle 'running', the past particle 'driven', and the conjunction 'before', respectively. They are syntactically ambiguous in many ways but are intended to have a unique reading as the most plausible semantic interpretation.

These concern particularly spatiotemporal (or 4D) relations, which are essential for human-robot interaction in casual scenarios. They were carefully designed so that ordinary people could have picturesque mental images evoked as their meanings, uniquely or not. The rather complex sentences I-6, I-7 and I-8 were employed for comparing the performances of the natural language understanding system named Conversation Management System (Khummongkol and Yokota, 2016) and human subjects for a certain psychological experiment concerning natural language understanding based on mental image arranged to show the validity and potentiality of mental image directed semantic theory. Conversation management system is provided with an apparatus in order to determine the most plausible reading of an ambiguous text, as described in Chapter 14.

In the author's opinion, the three machine translation systems seemed to work fairly well for forward translation but not so well for backward translation. The defect in the backward is greatly due to the outputs of the forward being not suitable for the inputs to the backward. Therefore, if the systems are provided with more elaborated natural language generation processes for forward translation, backward translation can be improved to a certain extent (but maybe only very little). For example, generation of less ambiguous target language texts will help the systems at least as for the test sentences above. This treatment seems feasible for Systran and Excite but questionable for Google Translate because specific errors are hard to predict and fix for machine-learning based systems. Actually, Excite has been found to respond to the unexpected results of users' trials in a short time.

In general, machine translation systems are designed for expected fluent speakers of the target language. Therefore, a polysemic word in the target language is very convenient for machine translation if it can mean the very meaning of the corresponding source language word as well on such an assumption that human users as target language speakers can select the right meaning. For example, the Japanese word 'de(で)' is very convenient and employed in such miscellaneous ways as '公園(kouen=park)で' for '*in* the park', '道路(douro=road)で' for '*on* the road', 'ナイフ(naifu=knife)で' for '*with* the knife', '腹痛(hukutsuu=stomach ache)で' for '*because of* stomach ache' and so on in English-to-Japanese translation. This is also the case for the Japanese word 'no(の)', for example, as '道路の車(kuruma=car)' for 'the car *on* the road', as '池(ike=pond)のカエル (kaeru=frog)' for 'the frog *in* the pond', and as '壁(kabe=wall)の色(iro=color)' for 'the color *of* the wall'. It is very trivial that these polysemic words should require commonsense knowledge about the two words connected with them in order for disambiguation.

As easily imagined, such a technique will fail in back-translation if the machine translation systems cannot cope with the polysemic words. Actually, the experimental results this time did not show their good successes in this point. However, this is quite natural because their task domains are not limited and therefore it is difficult for them to treat polysemic words such as 'with' appropriately. How well do they cope with polysemic words and ambiguous constructions without deep understanding or semantic processing? Google must have worked according to the corpora and the learning algorithms implemented. On the other hand, Systran and Excite seemed to be designed to replace polysemic source language words with target language words as general or abstract as possible, at least between English and Japanese.

The details about the three systems' performances are presented in the following subsections, including the author's subjective comments, where the forward and backward translations are indicated by the headers 'F' and 'B', respectively.

### 2.2.1 Google Translate (https://translate.google.com/)

The system seemed to have fluency at a glance but such a performance was found to be superficial upon closer inspection. This can be because it consults bilingual corpora without appropriate linguistic processing.
The experimental results are as follows, with some comments.

(I-1) With the box, the boy follows the girl.

➔ (F-1g) 箱で、少年はその少女の後を追う。(hako de, shounen wa sono shoujo no ato wo ou)

➔ (B-1g) In the box, the boy follows the girl.

Comment:

F-1g reads 'by means of the box, the boy chases the girl'. The preposition 'with' was translated to the Japanese polysemic word 'de(で)' which came to distort the original meaning in the backward. That is, the translation 'hako de (箱で)' reads 'by means of the box'. The most plausible translation of 'with the box' in Japanese would be 'hako wo motte(箱を持って)' meaning 'carrying the box' as the underlined part of F-3g while another translation 'hako no aru (箱のある)' in F-2g would be less plausible. The concerned parts are underlined.

(I-2) The boy with the box follows the girl.

➔ (F-2g) 箱のある男の子は女の子の後ろにいる。(hako no aru otoko no ko wa onna no ko no ushiro ni iru)

➔ (B-2g) A boy with a box is behind a girl.

Comment:

F-2g is somewhat strange, reading 'The boy whom the box is at is behind the girl' at best. 'The boy's 'moving' implied by 'follows' is lost in F-2g and naturally in B-2g, too because the Japanese verb 'iru(いる)' usually corresponds with the English verb 'stay' or 'be' used for stationarity.

(I-3) The boy follows the girl with the box.

➔ (F-3g) その男の子は箱を持って少女の後を追う。(sono otoko no ko wa hako wo motte shoujo no ato wo ou)

➔ (B-3g) The boy keeps a box and chases after the girl.

Comment:

F-3g is a good Japanese sentence, which is interpreted as 'The boy holds the box and chases the girl'. The other reading of I-3 'the girl with the box goes and the boy goes after her' was not inherited by F-3g because it is unambiguous. The translation from I-3 into F-3g implies an abundance of the selected case of dependency in the training set.

(I-4) Bring me the candy on the desk which is very sweet.

➔ (F-4g)私は甘い机の上にキャンディーを持って来てください。(watashi wa amai tsukue no ue ni kyandii wo motte kite kudasai)

➔ (B-4g) Please bring candy on a sweet desk.

Comment:

F-4g is a rather strange Japanese sentence but possibly reads '(*As*) for me (=私は(watashi wa)), bring the candy *onto* the *sweet desk* (=甘い机(amai tsukue))'. That is, 'sweet desk' was statistically preferred to 'sweet candy' in the forward. In B-4g, '(as) for me (=私は(watashi wa))' was dropped which, perhaps, statistically had better not exist.

(I-5) Look at the car on the road which is running very fast.

➔ (F-5g) 非常に速く走っている道路の車を見てください。(hijouni hayaku hashitteiru douro no kuruma wo mite kudasai)

➔ (B-5g) Please look at the car on the road running very fast.

Comment:

F-5g and B-5g are fairly good translations, preserving the syntactic ambiguity of I-5 due to its relative clause. That is, F-5g and B-5g are syntactically ambiguous in the same way as I-5 while people can understand that the car is running on the road by using their commonsense knowledge.

(I-6) Tom was with the book in the bus running from Town to University.

➔ (F-6g)トムは町から大学まで運行していたバスの本を持っていました。(tomu wa machi kara daigaku made unkoushiteita basu no hon wo motte imashita)

➔ (B-6g) Tom had a book of buses operating from town to university.

Comment:

F-6g is ambiguous in another way, reading 'Tom held the book which is *concerning/in* the bus operating from the town to the university'. The most plausible reading would be 'Tom, holding the book, was in the bus which was running from Town to University. The polysemic Japanese word 'no(の)' spoiled F-6g and B-6g.

(I-7) Tom was with the book in the car driven from Town to University by Mary.

➔ (F-7g) トムは、メアリーの町から大学へと車で運ばれた本を持っていました。(tomu wa, meari no machi kara daigaku eto kuruma de hakobareta hon wo motte imashita)

➔ (B-7g) Tom had a book that was taken by car from the town of Mary to the university.

Comment:

F-7g reads 'Tom held the book carried from Mary's town to the university'. The system failed to relate 'Mary' to 'driven' in the forward and in natural turn so in the backward. By the way, the phrase 'メアリーの町(meari no machi)' in F-7g corresponds with the dependency of 'by Mary' on 'Town' but it would be almost impossible in rule-based machine translation because it inevitably crosses or interferes that of 'to University' on 'driven'.

(I-8) Tom kept the book in a box before he drove the car from Town to University with the box.

➔ (F-8g) トムはボックスを持ってタウンから大学まで車を運転する前に箱に書いていた。(tomu wa bokkusu wo motte taun kara daigaku made kuruma wo untensuru mae ni hako ni kaite ita)

➔ (B-8g) Tom had a box and wrote in the box before driving a car from town to college.

Comment:

F-8g is interpreted as 'Tom wrote something on the box before he, holding *Box*, drove from *Town* to the university'. The noun phrase 'the box' was translated as a proper noun 'Box'. More strangely, the noun 'book' was eliminated in the forward process. Maybe, the colloquial combination of 'keep' and 'book' was statistically preferred. It seems to succeed in anaphoric identification of 'he' with 'Tom' but, at any rate, it is a result of the statistics of certain bilingual corpora.

### *2.2.2 Systran (http://www.systranet.com/)*

The system seemed to lisp somewhat compared with Google Translate, which can be due to having no examples of translations to simulate but composing them based on the implemented linguistic rules. This makes it possible to create translation systems for languages that have no texts in common, or even no digitized data whatsoever.

The experimental results are as follows with some comments.

(I-1) With the box, the boy follows the girl.

➔ (F-1s) 箱によって、男の子は女の子に続く。(hako ni yotte, otoko no ko wa onna no ko ni tsuzuku)

➔ (B-1s) With the box, the boy follows to the girl.

Comment:

F-1s reads 'by means of the box, the boy follows the girl'. The underlined part 'hako(箱) ni(に) yotte(よって)' means 'by means of the box'. The English verb 'follows' was translated into the combination of the Japanese words of 'ni(に)' and 'tsuzuku(続く)'. In B-1s, the two Japanese words were translated into the preposition 'to' and the verb 'follow(s)', separately. This is due to ambiguous word segmentation in Japanese without definite word boundary marking.

(I-2) The boy with the box follows the girl.

➔ (F-2s) 箱を持つ男の子は女の子に続く。(hako wo motsu otoko no ko wa onna no ko ni tsuzuku)

➔ (B-2s) The boy who has the box follows to the girl.

Comment:

F-2s is interpreted as 'the boy holding the box follows the girl'. Both the translations are fairly good except 'follows to the girl' in B-2s.

(I-3) The boy follows the girl with the box.

➔ (F-3s) 男の子は箱によって女の子に続く。(otoko no ko wa hako ni yotte onna no ko ni tsuzuku)

➔ (B-3s) The boy follows to the girl with the box.

Comment:

F-3s reads 'by means of the box, the boy follows the girl'. The other reading 'the girl goes/is with the book and the boy goes after her' was neglected.

(I-4) Bring me the candy on the desk which is very sweet.

➔ (F-4s) 私に非常に甘い机のキャンデーを持って来なさい。(watashi ni hijouni amai tsukue no kyandii wo motte kinasai)

➔ (B-4s) It has the candy of the very sweet desk in me, do.

Comment:

F-4s is good, preserving the syntactic ambiguity of I-4 due to its relative clause by use of the Japanese polysemic word 'no(の)'. However, the possibility of 'sweet candy' was eliminated in B-4s which substantially differs from I-4.

(I-5) Look at the car on the road which is running very fast.

➔ (F-5s) 非常に速く動いている道の車を見なさい。(hijouni hayaku ugoiteiru michi no kuruma wo minasai)

➔ (B-5s) Look at the car of the road which is moving very fast.

Comment:

F-5s is a good translation. The system preserved the syntactic ambiguity of I-5 due to its relative clause by the polysemic word 'no(の)' but its translation into 'of' distorted B-5s.

(I-6) Tom was with the book in the bus running from Town to University.

➔ (F-6s) トムは町から大学に走るバスの本とあった。(tomu wa machi kara daigaku ni hashiru basu no hon to atta)

➔ (B-6s) the book of the bus which it runs to the university from town there was tomu.

Comment:

F-6s is fairly good but B-6s would not be understandable. The proper noun 'トム (tomu)' was translated into 'tomu' as is pronounced.

(I-7) Tom was with the book in the car driven from Town to University by Mary.

➔ (F-7s) トムはメリーが町から大学に運転した車の本とあった。(tomu wa meari ga machi kara daigaku ni untenshita kuruma no hon to atta)

➔ (B-7s) the book of the car which Mary drove to the university from town there was tomu.

Comment:

F-7s is fairly good but B-7s would not be understandable.

(I-8) Tom kept the book in a box before he drove the car from Town to University with the box.

➔ (F-8s) トムは箱で彼が町から箱が付いている大学に車を運転した前に本を保った。(tomu wa hako de kare ga machi kara hako ga tsuiteiru daigaku ni kuruma wo untenshita mae ni hon wo tamotta)

➔ (B-8s) tomu him before driving the car to the university where the box has been attached from town preserved the book in the box.

Comment:

Both of the translations would be hard to be understood.

### 2.2.3 Excite (https://www.excite.co.jp/)

The performances of the system are considerably similar to Systran's, which is very natural because of the similar methodology.

The experimental results are as follows with some comments.

(I-1) With the box, the boy follows the girl.

➔ (F-1e) ボックスで、少年は女の子について行く。(bokkusu de, shounen wa onna no ko ni stuite iku)

➔ (B-1e) A boy goes about a girl by a box.

Comment:

F-1e reads 'by means of the box, the boy follows the girl'. The preposition 'with' was changed into the Japanese polysemic word 'de(で)' in the forward and then into 'by' in the backward. The English verb 'follows' was translated into the Japanese expression 'ni-tsuite-iku(について行く)' but in the backward 'ni-tsuite(について)' and 'iku(行く)' were translated into the English word 'about' and 'go(es)', separately.

(I-2) The boy with the box follows the girl.

➔ (F-2e) ボックスを持つ少年は、女の子について行く。(bokkusu wo motsu shounen wa, onna no ko ni tsuite iku)

➔ (B-2e) A boy with a box goes about a girl.

Comment:

F-2e is good but B-2e is anomalous because of 'goes about the girl'.

(I-3) The boy follows the girl with the box.

➔ (F-3e) 少年はボックスで女の子について行く。(shounen wa bokkusu de onna no ko ni tsuite iku)

➔ (B-3e) A boy goes about a girl by a box.

Comment:

F-3e is interpreted as 'by means of the box, the boy follows the girl'. The phrase 'with the box' was changed into 'ボックスで(=using the box)' in the forward, and into 'by a box' in the backward. B-3e is anomalous because of 'goes about the girl'.

(I-4) Bring me the candy on the desk which is very sweet.

➔ (F-4e) 非常に甘い机のキャンディーを私に持って来なさい。(hijouni amai tsukue no kyandii wo watashi ni motte kinasai)

➔ (B-4e) Bring a very sweet candy from desk to me.

Comment:

F-4e is a good translation. F-4e preserved the syntactic ambiguity of I-4 using the polysemic word 'no(の)' but it was eliminated rightly in B-4e.

(I-5) Look at the car on the road which is running very fast.

➔ (F-5e) 非常に速く動いている道路で車を見なさい。(hijouni hayaku ugoite iru douro de kuruma wo minasai)

➔ (B-5e) See a car on the road moving very fast.

Comment:

F-5e reads that the road is moving very fast, that is, the syntactic ambiguity of I-5 was wrongly eliminated. B-5e is syntactically ambiguous again due to the phrase 'moving very fast'.

(I-6) Tom was with the book in the bus running from Town to University.

➔ (F-6e) トムは、町から大学に走っているバスの本により持たれていた。(tomu wa, machi kara daigaku ni hashitteiru basu no hon ni yori motareteita)

➔ (B-6e) Tom was had by a book of the bus which runs to a university from a town.

Comment:

Neither of the two translations would be understandable, which is greatly due to 'hon ni yori motarete ita (本により持たれていた)' in F-6e for 'with the book'. This Japanese expression reads that Tom was held by the book and, in turn, it was strangely translated back to 'Tom was had by a book' in B-6e.

(I-7) Tom was with the book in the car driven from Town to University by Mary.

➔ (F-7e) トムは、町から大学にメアリーにより運転された車の本により持たれていた。(tomu wa, machi kara daigaku ni meari ni yori untensareta kuruma no hon ni yori motareteita)

➔ (B-7e) Tom was had by a book of the car driven by Mary in a university from a town.

Comment:

Neither of the two translations would be understandable. This is greatly due to the polysemic word 'ni(に)' in F-7e for 'to' in I-7, which was translated back to 'in' in B-7e.

(I-8) Tom kept the book in a box before he drove the car from Town to University with the box.

➔ (F-8e) 彼がボックスで町から大学に車を運転する前に、トムは本をボックスに保持した。(kare ga bokkusu de machi kara daigaku ni kuruma wo untensuru mae ni, tomu wa hon wo bokkusu ni hojishita)

→ (B-8e) before he drove a car in a university from a town by a box, Tom maintained a book in a box.

Comment:

Neither of the two translations would be understandable. This is greatly due to the underlined parts. The anaphoric identification of 'he' with 'Tom' seems to have been neglected.

## 2.3 Toward Mental Imagery-based Natural Language Understanding Beyond Conventional Natural Language Processing

Considering human cognitive process of viable translation, machine translation should be essentially cross-language paraphrasing via natural language understanding, preserving the meaning of the input source language text, probably without serious consideration of stylistic qualities of the generated target language texts. That is, machine translation should be one of the phases of natural language understanding, such that it interprets the source language text in a certain knowledge representation language and then generates the target language text from the interpretation. The core of such machine translation is naturally the knowledge representation language employed for the natural language understanding system.

So, what kinds of semantic processing should be required for natural language understanding in association with machine translation? Katz and Fodor (Katz and Fodor, 1963) presented the first analytical issue with human semantic processing ability. They claimed after their own experiences that people (more specifically, fluent speakers) can at least detect in a text or between texts such semantic properties or relations as follows and presented a model of disambiguation process called 'selection restriction' employing lexical information roughly specified by semantic markers and distinguishers in English.

a) semantic ambiguity
b) semantic anomaly
c) paraphrase relation (i.e., semantic identity between different expressions)

To our best knowledge, there has been no systematic implementation of these functions reported in any natural language processing or natural language understanding systems other than the work by Yokota and his comrades (Yokota, 2005; Khummongkol and Yokota, 2016). Among them, the most essential for natural language understanding is to detect paraphrase relation because the other two are possible if it is possible to determine equality (or inequality) between knowledge representations (or semantic representations) of different natural language expressions. As easily understood, the quality of this function depends on the capability of the adopted knowledge representation language to normalize knowledge representations, that is, to assign one knowledge representation to the same meanings. However, reflecting our psychological experiences in natural language understanding, we utilize tacit or explicit knowledge associated to the words or so involved in order to process an natural language expression semantically. This is also the case for natural language understanding systems. That is, they should be inevitably provided with knowledge good enough for the purpose, for example, lexical and ontological knowledge, computably formalized in knowledge representation language.

In general, a natural language understanding system is designed to map natural language expression into knowledge representation via certain grammatical description. Consider the test sentences I-1, I-2 and I-3. According to English grammar, I-1 is not syntactically ambiguous but I-2 and I-3 are so in two ways due to the prepositional phrase 'with the box'. Their grammatical descriptions can be given in dependency structure as D-1, as D-21 or D-22, and as D-31 or D-32, respectively, with the only difference at the dependency of the prepositional phrase 'with the box'. Viewed from natural language understanding, one of the problems is to identify the person who is carrying the box based on their corresponding meaning representations in knowledge representation language. Here, the prepositional phrase can depend on (or modify) the verb phrase 'follows' (D-1, D-22, D-31) or the noun phrases 'the boy' (D-21) or 'the girl' (D-32). It is naturally considered that only in the last case (D-32), the carrier of 'the box' is 'the girl' and otherwise, 'the boy'. Therefore, as for I-3, the system is to generate two different knowledge representations and detect semantic ambiguity. On the other hand, I-1 and I-2 are to be interpreted into the same knowledge representation and detected as paraphrases of each other in spite of their syntactic difference.

(D-1) follows(boy(the) girl(the) with(box(the))) ('with the box' depending on 'follows')

(D-21) follows(boy(the with(box(the))) girl(the)) ('with the box' depending on 'the boy')

(D-22) follows(boy(the) girl(the) with(box(the))) (=D-1)

(D-31) follows(boy(the) girl(the) with(box(the))) (=D-1)

(D-32) follows(boy(the) girl(the with(box(the)))) ('with the box' depending on 'the girl')

As for semantic anomaly, consider I-4 and I-5. If the systems are provided with lexical or ontological knowledge in the level of common sense, they will be able to detect semantic anomalies of 'sweet desk' and 'road running very fast' and thereby dissolve their syntactic ambiguities due to the relative clauses involved.

Now, analyze the processing results of the three machine translation systems based on the discussion above and it may be concluded that Excite seemed to employ selection restriction for local disambiguation (e.g., B-4e) but not for global (e.g., F-5e). That is, their qualities in semantic processing, such as disambiguation, are yet on the way to the level of human professional translators based on full understanding of the source language texts and good knowledge of the target language including commonsense knowledge. For example, it is not so easy for the machines to disambiguate the English sentence I-4, a rather simple expression, by identifying what is *sweet* because it is almost impossible for them to know every attribute of everything. This is the case for artificial inlelligence approaches to knowledge acquisition exclusively from existing text data, whether automatic or not, while people acquire such information almost subconsciously or automatically through all the senses as well. Therefore, lexical databases, such as WordNet (Miller et al., 1990), have been developed with the intention of systematic and efficient disambiguation or so by utilizing semantic relations tagged among words as tokens, for example, hyponym and hypernym, without defining individual word meanings explicitly. However, judging from the test results of the machine translation systems, without natural language understanding based on lexical and ontological knowledge in commonsense level, it would be difficult to translate relatively intricate source language texts, such as I-6–I-8, into readable target language texts. Distinguished from this approach to computational semantics, the knowledge representation language $\boldsymbol{L}_{md}$ in mental image directed semantic theory is designed to formalize individual word meanings and world knowledge based on the mental image model for predicate logic, and semantic relations such as hypernym and hyponym among them are considered to be deducible from them if necessary.

The concept of selection restriction is also very important for disambiguation in natural language understanding systems. More generally, however, they should be provided with a certain apparatus to assign 'semantic plausibility' to each of the readings in knowledge representation language of an input text on a continuous scale, where the lowest value should correspond with 'semantic anomaly'. For example, among the partial readings of I-7 involving the phrase 'by Mary', the semantic plausibility of 'the car was driven by Mary' would be greater than that of 'University was by Mary' or 'Town was by Mary', which agrees with commonsense knowledge and the results of our psychological experiments. Of course, this case can be resolved simply by capturing long-distance dependencies or the underlying predicate-argument structure (Baldwin et al., 2007). For another example concerning I-7, the part 'Tom was with the book in the bus' is syntactically ambiguous in two ways due to the prepositional phrase 'in the bus'. However, they should have the same semantic plausibility assigned to them because they are paraphrases each other. Actually, this is the case for conversation management system, which is because of the definition of the meaning of '*x* with *y*' in $\boldsymbol{L}_{md}$, reading that *x* causes *y* to be together, exclusively for the discourse domain of the physical world.

Table 2-1 shows the questions about I-6–I-8 and the answers by conversation management system which agreed with those of the human subjects very well. The human subjects were native speakers of Japanese, English, Chinese, or Thai. They were asked to sketch their mental images evoked by the stimulus sentences and to answer questions about them. Some disagreements were detected among all the participants (including conversation management system) and were found due to slight differences in their own word concepts. For example, one human

**Table 2-1.** Questions about I-6–I-8 and answers by conversation management system.

| Questions about I-6 (Tom was with the book in the bus running from Town to University). | Answers by conversation management system |
|---|---|
| Q1: What ran? | A1: bus |
| Q2: What was in the bus? | A2: Tom, book |
| Q3: What traveled from Town to University? | A3: Tom, bus, book |
| Q4: Did the bus carry Tom from Town to University? | A4: yes |
| Q5: Did the bus move Tom from Town to University? | A5: yes |
| Q6: Did the bus carry the book from Town to University? | A6: yes |
| Q7: Did the bus move the book from Town to University? | A7: yes |
| Q8: Did Tom carry the book from Town to University? | A8: yes |
| Q9: Did Tom move the book from Town to University? | A9: yes |
| Questions about I-7 (Tom was with the book in the car driven from Town to University by Mary.) | Answers by conversation management system |
| Q1: What was driven? | A1: car |
| Q2: What was in the car? | A2: Mary, Tom, book |
| Q3: What traveled from Town to University? | A3: Mary, Tom, book, car |
| Q4: Did the car carry Tom from Town to University? | A4: yes |
| Q5: Did the car move Tom from Town to University? | A5: yes |
| Q6: Did the car carry the book from Town to University? | A6: yes |
| Q7: Did the car move the book from Town to University? | A7: yes |
| Q8: Did Tom carry the book from Town to University? | A8: yes |
| Q9: Did Tom move the book from Town to University? | A9: yes |
| Q10: Did Mary carry the car from Town to University? | A10: yes |
| Q11: Did Mary carry the book from Town to University? | A11: yes |
| Q12: Did Mary carry Tom from Town to University? | A12: yes |
| Questions about I-8 (Tom kept the book in a box before he drove the car from Town to University with the box). | Answers by conversation management system |
| Q1: What traveled from Town to University? | A1: Tom, book, box, car |
| Q2: Did the car carry Tom from Town to University? | A2: yes |
| Q3: Did the car carry the box from Town to University? | A3: yes |
| Q4: Did the car carry the book from Town to University? | A4: yes |
| Q5: Did Tom carry the car from Town to University? | A5: yes |
| Q6: Did Tom carry the box from Town to University? | A6: yes |
| Q7: Did Tom carry the book from Town to University? | A7: yes |
| Q8: Did the box carry the book from Town to University? | A8: no |

subject answered only 'Tom' suitable for Q3 of I-6 because his concept of the verb *travel* was limitedly applicable to human agents who move from one place to another (for a special purpose such as sightseeing). That is, as detailed later, the psychological experiment has revealed that conversation management system can simulate human natural language understanding based on mental imagery with a good accuracy and that it has a good potential to be a paraphraser based on natural language understanding.

# 3

# Fundamentals for Robotic Natural Language Understanding

A language is a sign system in considering a vast range of communication forms, such as animal communication and man-machine communication. Viewed from natural language processing as man-machine communication interface, a natural language can be narrowly defined as a *symbol* system without any mathematically fixed prescription about its construction and meaning-making but with enormous usages by ordinary people who have learned it from its previous usages, so called 'corpus', indispensable to stochastic or statistic natural language processing nowadays. Therefore, natural language is essentially vague and ambiguous in every aspect of linguistics as a science. How can people manage their significant communication through natural language, which is rather irresponsible from a scientific viewpoint? It is considered that such a human performance should be derived from human competence of natural language understanding based on mental image which enables people to simulate and evaluate the meaning of an natural language expression as a virtual reality. Here, such a human competence is called mental-image based understanding, to be detailed in the later chapters. This chapter presents the essential requirements for natural language understanding ability of cognitive robots.

## 3.1 Natural Language Understanding in Accordance with Semiotics

According to semiotics, a general philosophical theory of signs and symbols, a sign system poses at least three theoretical departments to study as follows (Morris, 1938).

- **Syntax:** Theory on relations among signs in formal structures, that is, about how and which linguistic elements (as words) are combined to form constituents (as phrases or clauses) as the part of grammar (construed broadly to include phonology, phonemics, morphology and syntax).

- **Semantics:** Theory on relation between signs and the things to which they refer or their meaning as lexical knowledge, concerning denotation, extension, naming and truth.
- **Pragmatics:** Theory on relation between signs or linguistic expressions and their users, that is, about what in the actual situations the users intend based on a certain world knowledge.

The definitions of semantics and pragmatics are yet too vague for scholars to focus their efforts on because of a lack of objective data about semantic or pragmatic representation in human brains, namely, 'internal representation' (of meaning or knowledge). Therefore, among the three departments, syntax has been studied most successfully, resulting in the prescription of structures in the language, namely, the grammar, including its formalization for computation (Chomsky, 1956; Hays, 1967; Woods, 1970). The grammar is the first aid for people to acquire novel knowledge or information from others through language. For example, consider entities *X* and *Y* unknown to a native English speaker who is informed of them in such a context as '*X* eats *Y*'. This structure would make him know that *X* can be an animal and that *Y* can be a biological substance, distinctively from what people know by the structure '*Y* eats *X*'. Any grammar, however, is an artificial device *a posteriori* intended to rule the structures in the language, it is yet a hypothesis induced from a limited set of usages and is destined to ever evolve in order to defuse its exceptions. That is, there is no fixed theory in semiotics. Actually, ordinary people often use their own language rather ungrammatically in their casual scenarios, keeping their communication meaningful enough. This is because people have a very powerful semantic and pragmatic processing capability, enforced with general knowledge about the world, compared to what is called 'robust natural language understanding' in the field of artificial intelligence. Therefore, without any doubt, the goal of natural language understanding is to transplant such a human competence into computers, and its essence is considered to be a systematic methodology for formulation and computation of mental imagery, namely, internal representation of thought in humans.

Figure 3-1 shows the processing flow of natural language understanding considered in accordance with semiotics. The initial input is a sequence of symbols (already recognized and encoded for computation) and the main stream to the final output consists of symbol manipulations. In the case of a not unique solution, namely ambiguity, the result in each level of syntactic, semantic, and pragmatic processing is designed to be passed to the next processing level. This symbolic approach does not refuse any incorporation with statistical, connectionist, or other ones. A natural language understanding system is destined to respond in a certain style or protocol in order to show the evidence of its valid understanding of the input natural language expression. The final output can be varied according to the plan of what kinds of responses are to be synthesized, for example, paraphrase, translation in another language, answer to question, cross-media translation into still picture, animation, robotic action and so on.

The core of natural language understanding as the goal of natural language processing is the systematic methodology of knowledge representation and its computation for semantic or pragmatic understanding. Here, 'pragmatic understanding' means the total flow of natural language understanding in Fig. 3-1 and that without pragmatic analysis is called 'semantic understanding', especially.

So far, conventional approaches have proposed no clear definition of meaning itself and therefore no universal scheme for its computable formulation. That is, all conventional achievements that concern semantic and pragmatic processing, i.e., knowledge representation and computation, have been inevitably task-domain or goal oriented (e.g., Winograd, 1972; Schank and Abelson, 1977). In turn, this is also the case for syntactic analysis because it should output grammatical description optimized for semantic analysis to map natural language expression into knowledge representation. Each processing for syntactic, semantic and pragmatic analysis is to evaluate well-formed-ness of its input in each aspect and to infer its plausible results, so as to establish a single semantic or pragmatic reading (i.e., understanding result) as knowledge representation for the input natural language expression at last, namely, disambiguation.

Katz and Fodor asserted that synchronic linguistic description (a description of a natural language) minus **grammar** equals semantics (Katz and Fodor, 1963). This equation implies that they do not discern pragmatics from semantics, and simultaneously leads to degeneration of natural language expression analysis shown in Fig. 3-1 to that shown in Fig. 3-2. Actually, viewed from natural language understanding independent of robotic actions, such discrimination is not so important because both can be equally considered based on knowledge representation and computation without requirement of actual reference to or interaction with the external world. For example, consider S3-1 about its syntactic ambiguity, namely, about which is sweet, coffee or table. More probably, the answer is coffee but it is quite questionable whether this resolution is ascribable to lexical knowledge or world knowledge.

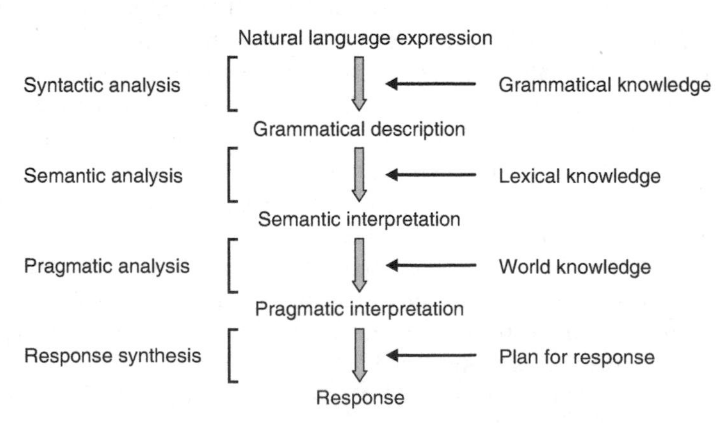


**Fig. 3-1.** Processing flow of natural language understanding for robots.

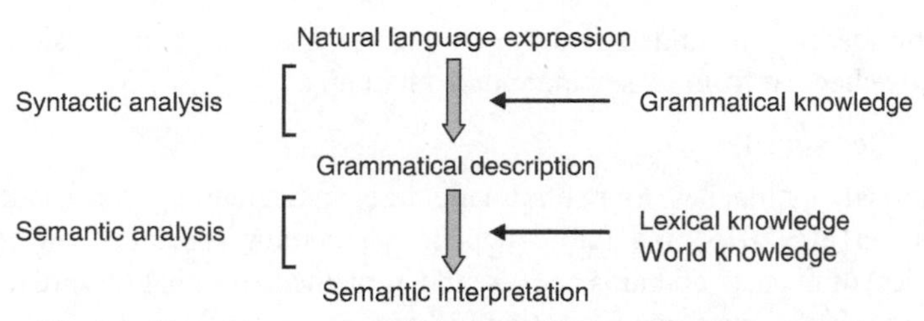


**Fig. 3-2.** Natural language expression analysis intended by Katz and Fodor.

(S3-1) There is a cup of coffee on the table which is very sweet.

However, viewed from natural language understanding for cognitive robots, semantics and pragmatics must be clearly distinguished in order to make an efficient decision on whether or not the input natural language expression should need pragmatic analysis for actual interaction with the external world as well as semantic analysis. For intuitive comprehension, suppose such a command as S3-2 is given by a human and imagine its understanding process by a robot as depicted in Fig. 3-3.

(S3-2) "Find a colorful box."

For more complicated example, consider S3-3, to know that this verbal command should require the robot to evaluate it semantically and pragmatically, namely, to resolve its syntactic ambiguity, anchor each word to its referent in the external world, and finally take the referent of 'soft and sweet candy' to the person referred to by 'me'.

(S3-3) "Bring me the candy on the table which is soft and sweet."

For another example, consider S3-4. People want their robot companion to stop when exclaiming so at the crossing. As easily understood by this example,

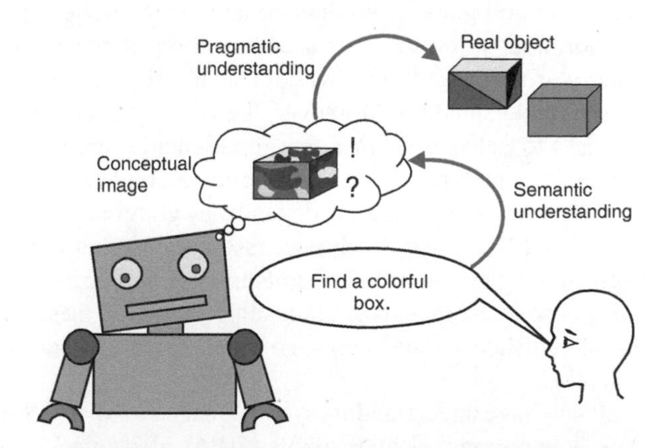


**Fig. 3-3.** Semantic and pragmatic understanding by a robot.

Color version at the end of the book

pragmatic reading (or understanding) of a natural language expression can be drastically changed from its semantic understanding.

(S3-4) "Red signal!"

Conventional approaches to natural language understanding for robots (e.g., Wächtera et al., 2018) inevitably depend on certain special knowledge (or pragmatics) of the task domains because the implemented semantics are too naïve to concern robotic sensation and action in general.

However, it is very important to provide robots with appropriate semantics. For example, consider S3-5–9. We can immediately discriminate which actions stated in them are feasible or not in the real world. These kinds of understandings are undoubtedly thanks to our certain commonsense knowledge of word meanings, namely, semantics. This is also the case for disambiguation of S3-10. Such semantic processing is very significant in the sense of preventing robots from wasting their energy for actions unfeasible at any cost or for truths unnecessary to prove in the real world, namely, without any pragmatic validation.

(S3-5) Catch the planet through the window.

(S3-6) Catch the insect through the window.

(S3-7) Travel to the city in the house.

(S3-8) Walk in the cupboard.

(S3-9) Walk in the forest.

(S3-10) See the moon in the room.

## 3.2 Syntactic Analysis

Traditional grammars are considered to determine a finite set of rules for well-formed structures or expressions in a language. In addition to this view, a grammar should be a good guide for bidirectional translation between a natural language expression and its meaning as a composition of the word meanings involved. In other words, any grammar without such a role should be of no use for natural language understanding. Moreover, the author's experiences have led to such a belief that any scheme for meaning representation in natural language understanding should be construed in a systematic correspondence with traditional grammars, especially, phrase structure and dependency grammars (conventionally abbreviated as PSG and DG, respectively) because it seems that phrase structure grammar and dependency grammar have something properly reflecting people's intuitive operations between language and meaning. Actually, these two are very helpful for him to translate natural language expressions and ideas as mental images into each other.

In our natural language understanding systems, such as IMAGES-M (Yokota, 2005) and CMS (Khummongkol and Yokota, 2016), dependency grammar is employed for grammatical description of a natural language expression as a set of inter-word dependencies to be systematically mapped into the knowledge representation as its semantic interpretation. There, phrase structure grammar is applied in advance to dependency grammar in order to restrict the scope of a dependency rule. That is, the final dependency structure forms a tree structure but has no phrasal nodes. Dependency structures are well suited for the analysis of languages with free word order, such as Japanese, Czech, and Slovak. In the Japanese language (to be detailed in Chapter 13), the postpositional particle, called *Joshi* (or more exactly *Kaku-joshi* (case-indicating *Joshi*)), plays the most important role in determining the root (or verb-centered) dependency structure of a clause. It is placed immediately after a noun and works like an English preposition.

### 3.2.1 Phrase structure grammar

Such definitions of phrases as PR-1 to PR-6 are employed in our natural language understanding systems, where the special symbols * and + represent an integer ≥ 0 and an integer ≥ 1, denoting the repetition times, respectively. In PR-2 for a noun phrase (NP), the adjective phrase (AJP) involved is optional, which is denoted by the pair of parentheses (see Table 3-1 for abbreviations of parts of speech and phrases of English).

(PR-1) AJP⇔ $Ad^*Aj^+$ (e.g., very much good polite)

(PR-2) NP⇔ $Det^*(AJP)N^+$ (e.g., all the very much good polite people communities)

(PR-3) VP⇔ $Ax^*V^+$ (e.g., can have been going)

(PR-4) PP⇔Pr NP (e.g., in the very big house)

(PR-5) ADP⇔ $Ad^+$ (e.g., very much fast)

(PR-6) COP⇔ $Ad^*Co$ (e.g., just before)

**Table 3-1.** List of parts of speech and phrases of English, and their abbreviations.

| Part of speech (abbreviation) | Phrase (abbreviation) |
|---|---|
| Adjective (Aj) | Adjective phrase (AJP) |
| Adverb (Ad) | Adverb phrase (ADP) |
| Auxiliary verb (Ax) | Noun phrase (NP) |
| Conjunctive (Co) | Prepositional phrase (PP) |
| Determiner (Det) | Verb phrase (VP) |
| Noun (N) | Conjunctive phrase (COP) |
| Preposition (Pr) | |
| Verb (V) | |

### 3.2.2 Dependency grammar

The dependency rules DR-1 to DR-15 for English below are part of those employed in our natural language understanding systems.

| | | |
|---|---|---|
| (DR-1) | Pr→N | (e.g., on the hill) |
| (DR-2) | Det←N | (e.g., the hill) |
| (DR-3) | Aj←N | (e.g., red book) |
| (DR-4) | Ad←Aj | (e.g., very large) |
| (DR-5) | Ad←Ad | (e.g., very much grateful) |
| (DR-6) | Ad←Aj | (e.g., very beautiful) |
| (DR-7) | Ax→V | (e.g., can do) |
| (DR-8) | V→V | (e.g., have been going) |
| (DR-9) | NP→PP | (e.g., the UFO on the hill) |
| (DR-10) | NP←VP | (e.g., Tom followed Jack) |
| (DR-11) | VP→NP | (e.g., Tom followed Jack) |
| (DR-12) | VP→PP | (e.g., Tom sat on the chair) |
| (DR-13) | PP←VP | (e.g., Tom with the stick ran) |
| (DR-14) | COP→VP | (e.g., Tom walked immediately after Jim ran) |
| (DR-15) | VP←COP | (e.g., Tom walked immediately after Jim ran) |

In these rules, Y←X and X→Y mean that the head X governs the dependent Y forward and backward, respectively, and both the cases are to be formulated as (3-1) for computation. A head word (X), such as verb appearing in a natural language expression, often takes multiple dependent words ($Y_1$ $Y_2$ … $Y_n$). Therefore, this should be more generally formulated as (3-2).

$$X(Y) \qquad (3\text{-}1)$$

$$X(Y_1\ Y_2 \ldots Y_n) \quad (n \geq 1) \qquad (3\text{-}2)$$

The rules DR-1 to DR-13 are valid within a clause while DR-14 and DR-15 are for inter-clause dependency. Among intra-clause rules, DR-1 to DR-8 are for intra-phrase and DR-9 to DR-13 are for inter-phrase. The intra-phrase rule DR-1, for example, reads that a preposition can be a head of a noun only within its own prepositional phrase. On the other hand, each of DR-9 to DR-13 for inter-phrase dependencies actually denotes the relation between the head words in either phrase, which is also the case for the inter-clause rules, DR-14 and DR-15.

In general, a dependency structure of a text is to be given as a nesting set of the units in the form of (3-2). For example, the grammatical description (i.e., dependency structure) of S3-11 is given as (3-3) by employing the dependency rules above.

(S3-11) Tom drove with the very pretty girl quite fast before he drank wine.

before (drove(Tom with(girl(the pretty(very))) fast(quite)) drank(he wine)) (3-3)

### 3.2.3 Sentence and discourse

As easily understood from the set of dependency rules above, the root of a dependency structure (i.e., X in (3-2)) is a verb for a simple sentence (or a clause) (dependency grammar condition-1) or a conjunction for a compound or complex sentence (dependency grammar condition-2). In other words, a sentence in English is defined as a grammatical constituent that satisfies either of these conditions in dependency grammar, especially here.

On the other hand, a discourse can be a maximal grammatical constituent which consists of a series of sentences combined with discourse connectives (Rouchota, 1996; Rysová and Rysová, 2018) including words or phrases, such as '(...,) too' and 'as follows', besides coordinate or subordinate conjunctions, such as 'and', 'or', and 'because'. A discourse connective is to be the root of the dependency structure of the discourse.

Judging from the author's investigation of several kinds of human languages, it is almost impossible to formalize all allowable expressions within a single grammar such as phrase structure grammar and dependency grammar. Therefore, a natural language processing system should be required to be provided with a certain mechanism to evaluate semantic plausibility of natural language expressions largely filtered through certain grammars, which is the case in our natural language understanding systems based on MIDST. For example, an English sentence containing multiple verbs without conjunctions (e.g., relative pronoun, present participle construction or so) is approximately treated as a set of multiple clauses combined with the conjunction 'and' in our system at the present stage.

### 3.2.4 Bidirectional mapping between natural language and knowledge representation via dependency structure

The grammatical rules presented above are part of those implemented in our natural language understanding systems. They are optimized for efficient mapping between natural language and $\boldsymbol{L}_{md}$, detailed in later chapters, however, note that they are almost the same as those accepted conventionally. It is considered that the most essential for natural language understanding is a systematic KRL, and any grammar is meaningless if it has no systematic mapping scheme between natural language expression and knowledge representation. A set of multiple dependency structures assigned to a text as its syntactic ambiguity are as well considered useful to approximate its semantic ambiguity because of almost one-to-one correspondence to its different semantic interpretations. However, we must note that dependency structure of itself is absolutely not semantic or knowledge representation.

In the author's own psychological experiences, an inter-word dependency can correspond to such a mental operation that the mental image evoked by the dependent modifies or instantiates a certain part of the head-word's mental image.

For example, S3-12 is syntactically and semantically ambiguous in six ways, (3-4)–(3-9), with one-to-one correspondence as shown in Fig. 3-4, although the combination B-D is conventionally to be eliminated because of their crossing each other that hinders us to imagine the scene being described. By the way, the subjects of our psychological experiment answered that they had only one image, either A-E or B-E, evoked. This fact implies that people have an intense association between the images of 'seeing' and 'telescope'.

(S3-12) Tom saw the UFO on the hill with the telescope.

(CASE A-C) saw(Tom UFO(the on(hill (the) with(telescope (the))))) (3-4)

(CASE A-D) saw(Tom UFO(the on(hill (the)) with(telescope (the)))) (3-5)

(CASE A-E) saw(Tom UFO(the on(hill (the))) with(telescope (the))) (3-6)

(CASE B-C) saw(Tom UFO(the) on(hill (the) with(telescope (the)))) (3-7)

(CASE B-D)* saw(Tom UFO(the with(telescope (the))) on(hill (the))) (3-8)

(CASE B-E) saw(Tom UFO(the) on(hill (the)) with(telescope (the))) (3-9)

However, for another example, S3-13 is syntactically ambiguous in two ways, (3-10) and (3-11), depicted in Fig. 3-5, due to DR-9 and DR-13, but its mental image as semantic interpretation is unique.

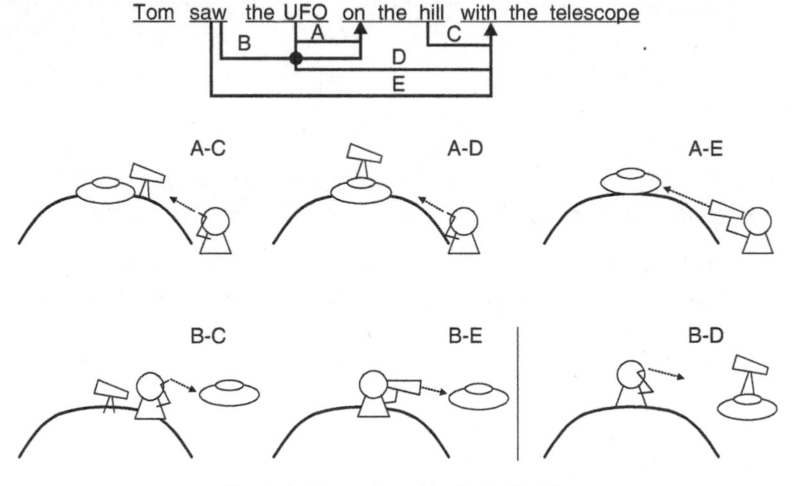


**Fig. 3-4.** Syntactic ambiguity in S3-12.

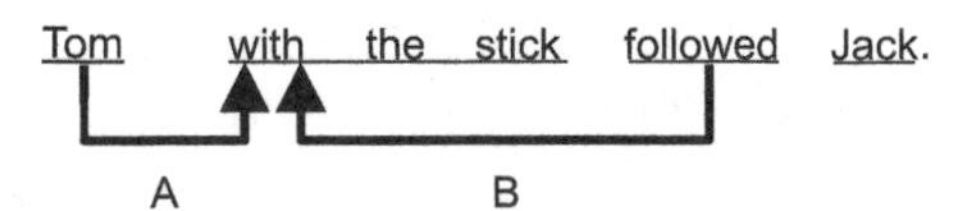


**Fig. 3-5.** Syntactic ambiguity in S3-13.

(S3-13) Tom with the stick followed Jack.

(CASE A) followed(Tom (with(stick(the))) Jack) (3-10)

(CASE B) followed(Tom with(stick(the)) Jack) (3-11)

## 3.3 Semantic Analysis and Pragmatic Analysis

Most remarkably to mention, Katz and Fodor presented the first analytical issue about human semantic processing competence. They claimed, based on their own experiences, that people (more exactly, fluent speakers) can at least detect among texts such semantic properties or relations as follows.

a) semantic ambiguity
b) semantic anomaly
c) paraphrase relation (i.e., semantic identity between different expressions)

Among these, the most essential for natural language understanding is to detect paraphrase relation because the others are possible if it is possible to determine equality (or inequality) between two knowledge representations. Schank's Conceptual Dependency theory (Schank, 1969) was intended to make the meaning independent of the words used in the input so that two sentences identical in meaning should have a single representation and applied to natural language understanding of limited scenes expressed in conceptual dependency called Scripts, such as restaurant and terrorism stories.

Anyhow, it must be noted that these detections essentially depend on the employed knowledge. For example, consider S3-14 and S3-15. Apparently, S3-14 is semantically ambiguous in two ways but S3-15 is probably not so because ordinary people know (or believe) that a house can be grounded in a hill but that the moon never touches a hill physically. That is, the two semantic readings of S3-14 are acceptable or normal but one reading for S3-15 is anomalous in commonsense level of the world knowledge (although such a metaphoric interpretation as S3-16 is possible). Notwithstanding, note that natural language understanding by people depends on their own knowledge about the language and the world concerned. That is, a natural language understanding methodology should provide a certain mechanism or framework to cope with such personal difference of mentality including knowledge. According to our psychological experiments, people are apt to interpret an ambiguous expression in a unique way, preferably led by their personal experiences, without becoming aware of such ambiguity instantly in the same way as with visual illusions. Wilks implemented such a people's tendency in natural language understanding as Preference Semantics (Wilks, 1972) in order to make machines select the best available interpretation of the input.

(S3-14) Tom saw the *house* on the hill.

(S3-15) Tom saw the *moon* on the hill.

(S3-16) To Tom, the moon *looked* grounded on the hill.

As easily understood, most metaphoric expressions are often semantically anomalous. For example, S3-17 describes human time sense with a profound meaning but it is absolute nonsense if evaluated in semantics based on scientific commonsense.

(S3-17) Time flies like an arrow.

As for paraphrase detection, any pair of literally same expressions can be semantically identical without considering the situations where they are uttered. Concerning situation dependency, for example, Tom described in S3-18 must be thrifty if he is an inhabitant of the desert with little water, but he must be wasteful if he is a resident in a town with abundant charge-free water supply. Apparently, this is not within the scope of semantic understanding but pragmatic understanding that depends on certain special world knowledge.

(S3-18) Tom spends money just like water.

The author and his comrades have already developed several intelligent systems with this kind of human competence for semantic analysis as the most basic function of natural language understanding (e.g., Yokota, 2005). To our best knowledge, our work is the first and only achievement of the requirements from Katz and Fodor. Already mentioned above, cognitive robots of practical use should be provided with pragmatic analysis well enough to interact with people in real environments. However, it is very crucial that they should be provided with good capability of semantic analysis to detect semantic ambiguity, semantic anomaly and paraphrase relation in order to prevent the robots from fruitless pragmatic analysis. Consider such a command as S3-19 given to a robot by a mischievous person. This expression is definitely nonsense in the real world and, therefore, the robot would come to work endlessly for such an impossible objective if it failed in detecting the semantic anomaly in the command. This kind of semantic anomaly can be detected rather easily based on the idea of selection restriction proposed by Katz and Fodor (Katz and Fodor, 1963). That is, in the case of S3-19, the *agent* of the action of swimming must be *animate*.

(S3-19) Find a swimming building.

## 3.4 Robust Natural Language Understanding

A natural language understanding system must be provided with a certain procedure against ill-formed inputs. It is quite easy for such a system to be designed to abandon any further processing for them but more desirable to anyhow abduct plausible interpretations from them. For example, consider S3-20 below.

(S3-20) 'Alaska, Tom gone.'

It is our knowledge of syntax that convinces us this expression is somewhat ungrammatical in English, and semantics that lead us to its plausible interpretation that someone named 'Tom' has gone to Alaska (in USA) based on word concepts

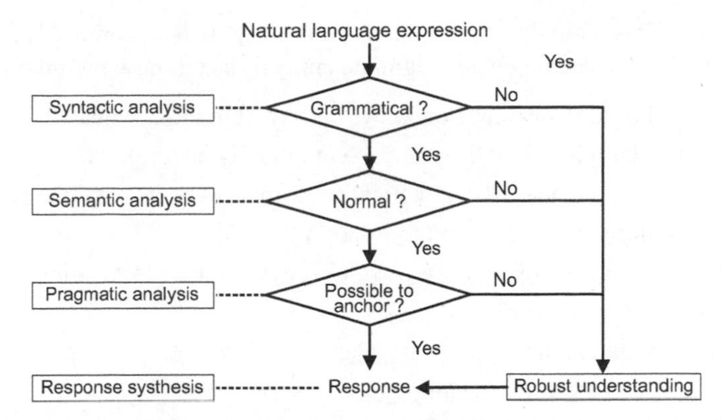


**Fig. 3-6.** Natural language understanding enforced with robust understanding module.

in commonsense level. It is pragmatics, however, that reveal the speaker's real intention, based on his special knowledge, that he wanted to tell his friend named 'Alaska' the death of their friend named 'Tom'. By the way, such an approach to natural language understanding that it should produce a plausible interpretation from an ill-formed input is called 'robust' natural language understanding. Considering the remarkable progress in speech recognition technology, speech understanding by computer in the near future will involve this kind of natural language understanding as its ideal style.

The natural language understanding system is intended to detect ill-formed inputs at every level of processing, namely, ungrammatical at syntactic analysis, anomalous at semantic analysis, or impossible (to anchor) at pragmatic analysis. It is robust natural language understanding that should correct ill-formed inputs to be well-formed and its general flow can be depicted as Fig. 3-6.

If all language users are assumed to try to be cooperative in order for further conversation, just like obeying the Grice's Conversational Maxims (Grice, 1975), but sometimes happen to fail due to mistakes by senders or receivers, any ill-formed natural language expression can be deemed to have some significant words lost or insignificant expressions filled on the way of communication and it can be turned well-formed by estimating them (so as to be well-formed). For example, S3-21 to S3-23 are examples of ill-formed inputs sensed at each level due to such happenings and S3-21' to S3-23' are the candidates for their original expressions, respectively, where the underlined words are assumed to have been lost. In principle, such semantic or pragmatic restorations are always possible by employing meta-level expressions for correction. For example, S3-23" to S3-23"" are such ultimate corrections for S3-23 which is contradictive to our knowledge as is. Furthermore, well-elaborated text or speech corpora specific to the discourse domains are very helpful for a robust natural language understanding system to retrieve candidates for correction based on morphological or phonological similarity in advance to semantic evaluation.

Non-lexical fillers, namely, strings out of vocabulary such as 'mmh…' and 'uh…' are to be passed to semantic and pragmatic analysis as unknown words.

(S3-21) Father mother went London yesterday. (Ungrammatical)

(S3-22) The path jumps at the point. (Semantically anomalous)

(S3-23) I am an elephant. (Pragmatically anomalous or impossible to anchor)

(S3-21') Father and mother went to London yesterday.

(S3-22') The path rises very acutely, so to speak, jumps at the point.

(S3-23') I am like an elephant.

(S3-23'') I am an elephant. It's a just joke.

(S3-23''') I am an elephant. I believe so.

(S3-23'''') I am an elephant. It's a lie.

The employed knowledge and the goal of the system are used to decide which parts are wrong or correct and how to correct wrong parts. For example, S3-24 could be corrected as S3-24' in commonsense knowledge and also as S3-24'' in Buddhist philosophy, based on morphological similarity between the words and semantical well-formed-ness.

(S3-24) Time collapses.

(S3-24') Time never collapses but lapses. (= Time lapses.)

(S3-24'') Time never collapses but thing. (= Thing collapses.)

By the way, most metaphoric expressions concerning time are often semantically anomalous. For example, S3-25 describes human time sense in the memory of a hard day but it is absolute nonsense if evaluated in semantics based on scientific commonsense.

(S3-25) It was the longest day yesterday.

The problems to be passed to robust natural language understanding are anomalous results in the three levels of processing, that is, syntactic, semantic, and pragmatic anomalies. The users of a natural language understanding system can be not only native speakers but also non-native speakers, prone to making mistakes in language use. To the utmost, however, they should be reasonable so that it can keep coherent dialogues with them and get necessary information from them for problem solving. Based on such an assumption, the best strategy for problem solving in natural language understanding by robots, namely, robust natural language understanding, must be question-answering with the reasonable users about the very problems. Nonetheless, a desirable natural language understanding system must be diligent in problem solving and kind to the users. For example, consider such a case that a user utters "Father mother Alaska go." This expression is apparently quite ungrammatical but a natural language understanding system should treat the user with such a thoughtful comment as "Do you mean that your father and mother went or will go to Alaska?" rather than such a flat one as "I cannot understand what you say."

Here is described 'robust' natural language understanding driven by semantic and pragmatic knowledge in order to understand ill-formed expressions as well as well-formed ones. Therefore, the author's efforts are to be concentrated on formalization and computation of its semantics and pragmatics.

For example, consider the very famous Japanese sentence S3-26 called 'unagi-bun (=eel-sentence)' whose equivalent in English could be such as S3-26'. Almost all Japanese-English translation programs in the market are apt to output such a sentence as S3-27 while the plausible translation should be such as S3-27' and S3-27'', reflecting Japanese cultural background as pragmatics, that is, most Japanese people like to eat eel and often employ such an expression as S3-26 in order to declare what they want to have.

(S3-26) Boku wa *unagi* da.

(S3-26') Me, *eel.*

(S3-27) I'm an *eel.*

(S3-27') I'd like to eat *eel.*

(S3-27'') I'm fond of *eel.*

## 3.5 Response Synthesis

We provided our natural language understanding systems with a protocol for response for each sentence type, such as declarative and interrogative, one by one.

At first, a piece of information (***I***) is formalized as a set of messages, (*m*'s) is formalized as (3-12).

$$\boldsymbol{I}=\{m_1, m_2, ..., m_n\} \quad (3\text{-}12)$$

In turn, a message (*m*) is defined in the expression (3-13), where *D, S, R* and *B* mean the duration, sender(s), receiver(s) and the body of the message, respectively.

$$m=(D, S, R, B) \quad (3\text{-}13)$$

The body (*B*) consists of the two elements shown in (3-14), where *E* and *T* mean the event referred to and the task requested or intended by the sender, respectively.

$$B=(E, T) \quad (3\text{-}14)$$

For example, consider each item of the message $m_0$: "Fetch me the milk from the fridge, Anna" uttered by *Taro* to his companion robot, *Anna*, during the time-interval [$t_1$, $t_2$]. This message is syntactically, semantically, and pragmatically analyzed and finally formulated as (3-15)–(3-18). Here, the formulations of $E_0$ and $T_0$ are informal but, actually, they are formalized in the knowledge representation language $\boldsymbol{L}_{md}$ so as to be computable in our systems, as detailed

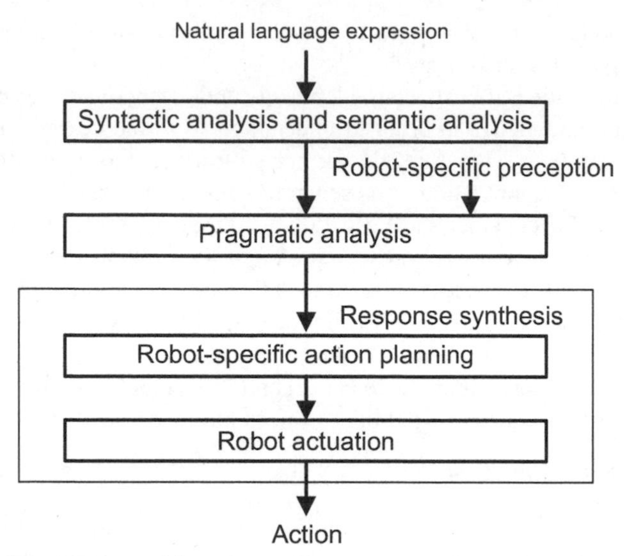


**Fig. 3-7.** General flow of natural language expression to robot action.

later. The Italicized nouns in (3-17) denote that they are successfully anchored to the real objects in the environment through Anna's machine-specific perception at pragmatic analysis, as shown in Fig. 3-7.

$$m_0=(D_0, S_0, R_0, B_0), B_0=(E_0, T_0) \quad (3\text{-}15)$$

$$D_0=[t_1, t_2], S_0=\text{“Taro”}, R_0=\text{“Anna”} \quad (3\text{-}16)$$

$$E_0=\text{“Anna FETCH Taro milk FROM fridge”} \quad (3\text{-}17)$$

$$T_0=\text{“realization of } E_0\text{”} \quad (3\text{-}18)$$

Response Synthesis department shown in Fig. 3-7 plans the action for $E_0$ specific to Anna's actuators and realizes it in the environment according to $T_0$. Meanwhile, during planning and execution of the action, various kinds of problems must be found and solved by coordinating sensors and effectors (=actors or actuators) (Khummongkol and Yokota, 2016), for example, obstruction avoidance.

There are orthodox correspondences between the kinds of tasks (*T*'s) and the types of sentences as shown in Table 3-2, which are very useful for computation. Pragmatically, every sentence type can be employed for another intention just like rhetorical question. For example, we do not need to answer such a question as 'Who knows?' when we are asked so because its answer should be 'Nobody knows.' For another example, when people utter 'The signal is red' at a crossing, they surely imply a special intention of caution. This unorthodox usage of declarative can be treated as omission of a certain expression such as 'Do not cross the road' or 'Stop'. For such cases, we are ready to provide the system with a certain module for intention estimation.

**Table 3-2.** Protocol about English sentence types and tasks in the natural language understanding system.*

| **Sentence type (Examples)** | **Task (*T*)** |
|---|---|
| Declarative (or exclamatory)<br>(It is ten o'clock now./ How great it is!) | To believe *E*. |
| Interrogative<br>([A] Is it ten o'clock now?<br>[B] What time is it now?) | [A] To reply whether *E* is true or false.<br>[B] To reply what makes *E* true. |
| Imperative<br>(Show me your watch.) | To realize *E*. |

*In principle, it is assumed that the receiver can respond to the sender in any way not limited to this protocol.

## 3.6 Syntax and Semantics of Discourse

Our communication usually forms a unit called *discourse*, consisting of multiple utterances. Therefore, ideally, a natural language understanding system should be a discourse understanding system. In linguistics, a discourse is defined as any unit of connected speech or writing longer than a sentence, and there are two types of words or phrases to characterize it, namely, discourse connectives and discourse markers (Cohen, 1984; Schiffrin, 1987).

A discourse connective (such as *and, however*, or *as follows*) has a function to express the semantic connection between the discourse units prior to and following it (Cohen, 1984; Rouchota, 1996; Rysová and Rysová, 2018). On the other hand, a discourse marker (such as *oh, well, like, so, okay, I mean*, and *you know*) (Schiffrin, 1987; Swan, 2005) is almost syntactically independent and has a pragmatic function or role to manage the flow of conversation without changing any semantic content (i.e., meaning) of the discourse. Therefore, here, discourse markers are excluded from grammatical or semantic description of a discourse.

Mental image directed semantic theory introduces a discourse as the maximal constituent defined as a series of sentences (or utterances) by (3-19), where a discourse $D$ is a series of sentences $S$ respectively containing a discourse connective $C_d$ and denoted as $S[C_d]$. In turn, the discourse connective $C_d$ is defined by (3-20) as a list of words or phrases, not limited to single conjunctions (i.e., *Co*), where the special symbol $\phi$ corresponds with the case of its omission or implicit establishment of relation (mostly of the conjunction *and*) which is often observed in actual scenarios, especially in writing, e.g., parataxis.

$$D \Leftrightarrow S^n[C_d] \Leftrightarrow S[C_d]\, S[C_d] \ldots S[C_d] \quad (n \geq 1) \tag{3-19}$$

$$C_d \Leftrightarrow \phi | Co | moreover | however | too | as\ follows | The\ reason\ is\ that | \ldots \tag{3-20}$$

In (3-19), the first sentence, as well as the following ones, can contain a discourse connective to link with some prior undescribed or unspoken context. For example, when you find your friend back from jogging, you can give her such an utterance as "*So* you're trying to keep fit" or "I should pay some attention to my fitness *too*" (Rouchota, 1996). Moreover, as easily understood, the definition of a discourse holds also for a single sentence consisting of multiple clauses or verbals, such as infinitives and present and past participles. That is, here, a sentence with multiple verbs or verbals is recognized and processed as a discourse or its equivalent.

Consider a discourse $D_0$, formulated as (3-21), consisting of two units $D_1$ and $D_2$ and a discourse connective $C_d$.

$$D_0=D_1\ C_d\ D_2 \tag{3-21}$$

In all the same way as the processing of a single sentence described in 3.2.4, $D_0$ is to be converted into the dependency structure (3-22) and in turn into the semantic representation in $\boldsymbol{L_{md}}$ (3-23). The processes to translate natural language into dependency structure and dependency structure into $\boldsymbol{L_{md}}$ are denoted as functions $f$ and $g$, respectively, and $f(D_i)$ and $g\,(f\,(D_i))$ represent the dependency structure and the $\boldsymbol{L_{md}}$ expression of $D_i$, respectively.

The root (i.e., head word) of the dependency structure of a discourse is to be a discourse connective, and, therefore, there are two inter-discourse (or inter-sentence) dependency rules, namely, DR-16 and DR-17, where $H_w$ denotes the head word of the dependent discourse unit. When the dependent discourse unit is a simple sentence, $H_w$ is a verb, and otherwise a discourse connective.

$$f(D)= C_d\,(\,f(D_1)\ f(D_2)) \tag{3-22}$$

$$g(f\,(D))= g(C_d\,((D_1)\ f(D_2))) \tag{3-23}$$

(DR-16) $H_w \leftarrow C_d$

(DR-17) $C_d \rightarrow H_w$

On the other hand, an English sentence containing multiple verbs without conjunctions (e.g., relative pronoun, present participle construction or so) is approximately treated as a set of multiple clauses connected with the conjunction (or discourse connective) *and* in our system. For example, consider several pairs of paraphrases, S3-28 and S3-28' to S3-31 and S3-31', where the underlined parts are required for anaphora.

(S3-28) Tom is in the bus (which is) running fast.

(S3-28') Tom is in the bus. *And*, it is running fast.

(S3-29) Tom is in the car (which is) driven fast by Jim.

(S3-29') Tom is in the car. *And*, it is driven fast by Jim.

(S3-30) Tom bought a house to live in.

(S3-30') Tom bought a house. *And*, Tom would/will live in it.

(S3-31) Tom likes dancing/to dance with Mary.

(S3-31') Tom likes a *matter*. *And*, *the matter* is that he dances with Mary.

It is a matter of course that the paraphrases above can require further elaboration. Especially, S3-31' can be considerably unnatural because the term 'matter' is technically employed as the hypernym of 'actuality' and 'reality'.

Anyway, the paraphrasing schemes above are formulated in (i)–(iii) as follows, where X and Y are a clause or its equivalent. The notation A⇒B⇒C implies translation from natural language text (A) into dependency structure of the paraphrase (B) and into $\boldsymbol{L}_{md}$ expression (C), although concrete description of semantic interpretation (i.e., B⇒C) is omitted in this section and to be detailed in the later chapters.

i) Construction of relatives

The scheme is given as (3-24), where *rel* and *w* represent a relative and its antecedent to be supplemented by anaphora, respectively.

$$\mathrm{X\ Y(rel)} \Rightarrow \mathrm{and}(\mathfrak{f}(\mathrm{X})\ \mathfrak{f}(\mathrm{Y(w)})) \Rightarrow \mathscr{g}(\mathfrak{f}(\mathrm{X})) \wedge \mathscr{g}(\mathfrak{f}(\mathrm{Y(w)}) \quad (3\text{-}24)$$

For example, 'Tom is in the car which he drives' is converted to the dependency structure (3-25).

and(is(Tom in(car(the))) drives(he w)) (3-25)

ii) Construction of verbals in adjective or adverb use

The scheme is formulated as (3-26) or (3-27), where *v* and *w* are a verbal and its omitted subject or so to be supplemented by anaphora, respectively.

$$\mathrm{X\ Y(v)} \Rightarrow \mathrm{and}(\mathfrak{f}(\mathrm{X})\, f(\mathrm{Y(v(w))})) \Rightarrow \mathscr{g}(\mathfrak{f}(\mathrm{X})) \wedge \mathscr{g}(\mathfrak{f}(\mathrm{Y(v(w))})) \quad (3\text{-}26)$$

$$\mathrm{X(v)\ Y} \Rightarrow \mathrm{and}(\mathfrak{f}(\mathrm{X(v(w))})\ \mathfrak{f}(\mathrm{Y})) \Rightarrow \mathscr{g}(\mathfrak{f}(\mathrm{X(v(w))})) \wedge \mathscr{g}(\mathfrak{f}(\mathrm{Y})) \quad (3\text{-}27)$$

For example, 'Tom is in the bus running fast' is converted into the dependency structure (3-28).

and(is(Tom in(bus(the))) running(w fast)) (3-28)

For another example, the dependency structure of 'Weeping, the baby drank the milk' is given as (3-29).
and(weeping(w) drank(baby(the) milk(the))) (3-29)

iii) Construction of verbals in noun use

The paraphrasing scheme is formulated as (3-30), where *v*, $w_A$ and $x_e$ are a verbal, its omitted subject or so to be supplemented by anaphora, and an extra-symbol for some matter, respectively. According to this scheme, the verbal construction is to be translated into the clausal construction Y'(v'($\mathrm{w_A}$)).

$$\mathrm{X(Y(v))} \Rightarrow \mathrm{and}(\mathfrak{f}(\mathrm{X(x_e)})\ \mathfrak{f}(\mathrm{x_e\ is\ that\ Y'(v'(w_A))})$$

$$\Rightarrow \mathscr{g}(\mathfrak{f}(\mathrm{X(x_e)})) \wedge \mathrm{x_e} = \mathscr{g}(\mathfrak{f}(\mathrm{Y'(v'(w_A))})) \quad (3\text{-}30)$$

For example, 'Tom likes swimming/to swim' is translated into the dependency structure (3-31).

and(likes(Tom $\mathrm{x_e}$) is($\mathrm{x_e}$ that(swims(Tom))) (3-31)

The paraphrases described above cannot conserve the clausal or verbal subordination in the original expressions. However, they can be appropriate enough for human-robot interaction through natural language in the physical world because the ostensive 4D scenes being described or referred to are the same. That is, the (non-syntactic) difference here between an original expression and its paraphrase is not semantic but pragmatic.

In mental image directed semantic theory, every discourse connective is to be explicitly given a meaning definition with a set of operation commands about how to connect the discourse units and Yokota (Yokota, 1999) has proposed a computational model of the mechanism to select a discourse connective based on the semantic contents of (prior and following) discourse units and speaker's (or writer's) belief (broadly including knowledge).

Consider the facts being described by sentence $S_1$ and sentence $S_2$ below.

$S_1$: It is raining.
$S_2$: Tom is staying at home.

When the speaker observes the facts and has such a belief (piece) as $S_3$, she can utter $S_4$.

$S_3$: If it is raining, then Tom is staying at home. $(=S_1 \supset S_2)$
$S_4$: Tom is staying at home *because* it is raining.

Her thinking process to make the utterance $S_4$ can be formalized as follows.

Firstly, her observation and belief are given as (3-32) and (3-33), respectively, where $\mathcal{B}$ is the total set of her belief pieces.

$$S_1 \wedge S_2 \tag{3-32}$$

$$S_1 \supset S_2 \in \mathcal{B} \tag{3-33}$$

Secondly, she tries confirming the belief (3-33) in such a process as (3-34). That is, the belief piece is equivalent to TRUE.

$$S_1 \wedge S_2 \wedge (S_1 \supset S_2) \leftrightarrow S_1 \wedge S_2 \wedge (\sim S_1 \vee S_2) \leftrightarrow S_1 \wedge S_2 \tag{3-34}$$

Finally, she comes to utter $S_4$ concluding that $S_1$ has caused $S_2$.

Therefore, the meaning definition of the conjunction *because* can be formalized as (3-35), where the underlined part is out of semantics but in pragmatics because it is specific to the speaker's belief. That is, *because* is equal to the logical AND within semantics.

$$Y \text{ because } X \Leftrightarrow X \wedge Y \wedge (\underline{X \supset Y \in \mathcal{B}}) \tag{3-35}$$

By the way, in the case of predictive inference based on the fact (or premise) *X*, at least the event referred to by the conclusion *Y* has not been observed yet.

For another example, consider the fact $S_5$ in place of $S_2$.

$S_5$: Tom is not staying at home. (=~$S_2$)

In this case, she can utter $S_6$ because she comes across contradiction between her observation (3-36) and the conclusion deduced from the fact $S_1$ and her belief $S_3$, as shown by (3-37).

$S_6$: It is raining *but* Tom is not staying at home.

$$S_1 \wedge S_5 \equiv S_1 \wedge \sim S_2 \quad (3\text{-}36)$$

$$S_1 \wedge \sim S_2 \wedge (S_1 \supset S_2) \leftrightarrow S_1 \wedge \sim S_2 \wedge (\sim S_1 \vee S_2) \leftrightarrow \text{FALSE} \quad (3\text{-}37)$$

And, therefore, the meaning of *but* can be defined as (3-38).

$$X \text{ but } \sim Y \Leftrightarrow X \wedge \sim Y \wedge (X \supset Y \in \mathcal{B}) \quad (3\text{-}38)$$

As easily understood, it is very important for a robot to discern the semantic part and the pragmatic part in a meaning definition because the latter is usually unobservable, unlike the former (especially concerning the physical world).

Furthermore, it is noticeable that, in general, the material implication statement $p \supset q$ ($\Leftrightarrow \sim p \vee q$) does not specify a causal relationship between $p$ and $q$. Therefore, the other two possible cases (3-39) and (3-40) also make (3-33) TRUE but the utterance $S_7$ and $S_8$ would sound strange and nonsensical, respectively.

$$\sim S_1 \wedge \sim S_2 \quad (3\text{-}39)$$

$$\sim S_1 \wedge S_2 \quad (3\text{-}40)$$

$S_7$: Tom is not staying at home *because* it is not raining.

$S_8$: Tom is staying at home *because* it is not raining.

If $S_7$ sounded good, the belief could be (3-41) implying that *Tom is staying at home if and only if it is raining.*

$$S_1 \equiv S_2 \in \mathcal{B} \quad (3\text{-}41)$$

On the other hand, with the same belief, such an utterance $S_9$ instead of $S_8$ could be allowed, where the meaning definition of *although* can be given by (3-42).

$S_9$: (To my surprise) Tom is staying at home *although* it is not raining.

$$Y \text{ although } \sim X \Leftrightarrow \sim X \wedge Y \wedge (X \equiv Y \in \mathcal{B}) \quad (3\text{-}42)$$

# 4
# Cognitive Essentials for Mental Image Directed Semantic Theory

This chapter introduces several ideas concerning human and robotic cognitions essential for mental image directed semantic theory.

## 4.1 Functional Model of the Human Mind

The author thinks that it is essential for cognitive robotics to consider how the human mind works. This section presents a functional model of the human mind, assumed to work based on mental image. Figure 4-1 shows the multi-agent mind model proposed here, much simpler than Minsky's (Minsky, 1986), consisting of Stimulus, Knowledge, Emotion, Integration, and Response Processing Agents in large. This is a functional model of the human central nervous system, consisting of the brain and the spine. These agents communicate with one another by exchanging and computing mental images (i.e., conceptual, perceptual, or hybrid images) represented in the formal language mental image description language (i.e., $L_{md}$). Their basic performances are as follows.

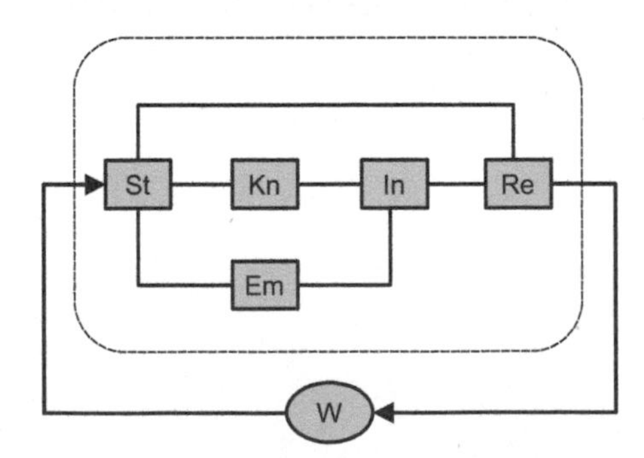


**Fig. 4-1.** Multiagent model of the human mind (St: Stimulus Processing Agent; Kn: Knowledge Processing Agent; Em: Emotion Processing Agent; In: Integration Processing Agent; Re: Response Processing Agent; W: World surrounding human mind, including his/her body).

1) **Stimulus Processing Agent** receives stimuli from the **world** and encodes them into mental images (i.e., encoded sensations) such as "*I sensed something soft.*" (if verbalized in English.)
2) **Knowledge Processing Agent** evaluates mental images received from the other agents based on its memory (e.g., knowledge), producing mental images such as "*I've recognized that it is a piece of marshmallow.*" The results are called *rational mental images.*
3) **Emotion Processing Agent** evaluates mental images received from the other agents based on its memory (e.g., instincts), producing mental images such as "*I like the food.*" The results are called *emotional mental images.*
4) **Integration Processing Agent** evaluates the outputs from **knowledge processing agent** and **emotion processing agent** in total and produces images such as "*Then, I'll take and eat it*" at decision about the response to the initial stimuli. When this agent is working, we are *think*ing about something.
5) **Response Processing Agent** converts mental images (i.e., encoded actions such as "*Take and eat it.*") received from the other agents into real actions upon the **world.**

A *performance* ***P*** against a *stimulus X* with a *result Y* at each agent can be formalized as a function by the equation (4-1).

$$Y=\mathbf{P}(X), \qquad (4\text{-}1)$$

where

*P*: A combination of ***Atomic Performances*** defined later in association with ***Attribute Spaces***;
*X*: A spatiotemporal distribution of stimuli from the **world** to **stimulus processing agent** or a mental image for another agent;
*Y*: A series of signals to drive actuators for **response processing agent** or a mental image for another agent.

For example, all the agents are to work during understanding information media, such as natural language, picture, music, gesture, etc., sometimes performing *Kansei* by tight collaboration of **knowledge processing agent** and **emotion processing agent** via **integration processing agent**, while **stimulus processing agent** and **response processing agent** are exclusively to work during *reflection. Kansei* evaluates non-scientific matters, such as art, music, natural scenery, etc., by Kansei words (e.g., heart-calming, fantastic, grotesque) and *Artificial Kansei,* namely *Kansei for robots,* is expected to play a part in artificial or robotic individuality (Tauchi et al., 2006; Shiraishi et al., 2006; Yokota et al., 2008).

A performance ***P*** is assumed as a function formed either consciously or unconsciously, or in other words, either with or without reasoning. In a conscious case, a set of atomic performances are to be selected and combined according to *X* by a meta-function, so called, ***Performance Selector*** assumed as ***Consciousness.***

On the contrary, in an unconscious case, such a performance as is associated most strongly with *X* is to be applied automatically as in the case of reflection.

It is well known that emotion in people can be affected by their world, namely, the **world** in Fig. 4-1. For example, their evaluation of live image of an object (i.e., image output from **stimulus processing agent**) expressed by such words as favorite, beautiful, tasty, etc., can vary depending on their emotional bias, such as hunger, depression, etc.

For example, *Kansei* is considered as one of **the collaborations between knowledge processing agent and emotion processing agent**. This has a more complicated phase than pure emotion originated from instinct or imprinting. For example, sweet jam may be nice on toast but not on pizza for certain people knowledgeable about these foods. For another example, people can be affected on their evaluation of an artwork by its creator's name, for example, Picasso. These are good examples of *Kansei* processing as emotional performance affected by knowledge in humans.

Therefore, *Kansei* can be defined as human emotion toward an object affected by its significance for them, so called, concept, including their intellectual pursuits, traditions, cultures, and so on concerning it. In this sense, *Kansei* is assumed to be reasonable among the people sharing such concepts, unlike pure emotion. These hypothetic considerations are formalized as (4-2) and (4-3).

$$I_P(\mathrm{x})=\boldsymbol{P}_E(\mathrm{S}(\mathrm{x})) \tag{4-2}$$

$$I_K(\mathrm{x})=\boldsymbol{P}_E(\mathrm{S}(\mathrm{x})\wedge \mathrm{O}(\mathrm{x})) = \boldsymbol{P}_E(\mathrm{S'}(\mathrm{x})) \tag{4-3}$$

where

$\boldsymbol{P}_E$(X): Performance of **emotion processing agent** for mental image *X*,

$I_P$(x): Mental image as pure emotion for object *x*,

$I_K$(x): Mental image as *Kansei* for object *x*,

S(x): Live image of object *x* from **stimulus processing agent,**

O(x): Concept of object *x* from **knowledge processing agent,**

S'(x): Unified image of live image and concept.

Figure 4-2 shows an example of *Kansei* processing in our mind model, where perceived, induced and inspired images correspond to S(x), S'(x) and $I_K$(x), respectively, while Fig. 4-3 is for pure emotion with $I_P$(x) as the inspired image.

These two inspired images are to be verbalized in **response processing agent** as 'Fragrant!' and 'Appetizing!', labeled in Fig. 4-3, respectively. The essential difference between them is assumed to reside in whether or not they are affected by O(x), namely, the concept of 'chocolate cream bread', inferred by **knowledge processing agent** from the shape and the smell. Whereas, pure emotion for an object can be a special case of *Kansei* processing without knowing or recognizing what it is.

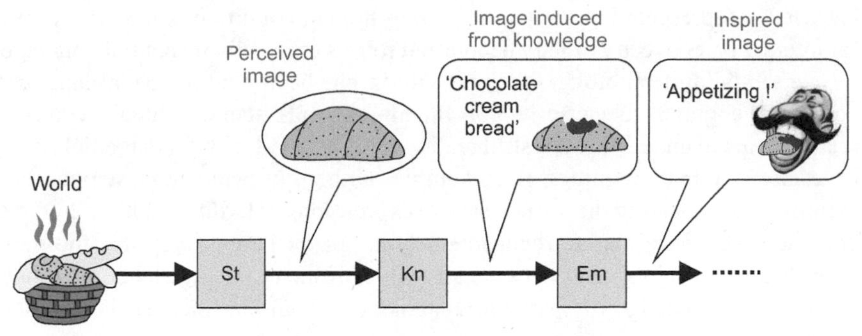


**Fig. 4-2.** Example of *Kansei* processing.

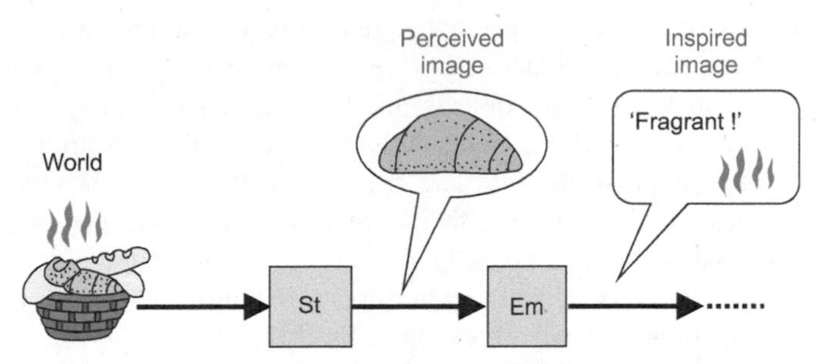


**Fig. 4-3.** Example of pure emotion.

## 4.2 Human Knowledge and Cognitive Propensities

Conventionally, knowledge representation is defined as the application of logic and ontology to the task of constructing computable models of some domain (Sowa, 2000). That is, ontological commitment as well as computation-awareness (or consciousness) is indispensable in designing a knowledge representation language. However, almost all such ontological commitments were quite objective, namely, without any reflection of human cognitive processes or products, although ordinary people live the greater part of their casual life based on their own subjective knowledge of the world, acquired through their inborn sensory systems. For example, in our daily life, it is much more convenient for us to know that the sun rises in the east and sets in the west than that the earth is spinning around the sun. Actually, according to the author's long-term study, it is very clear that people's natural concepts (i.e., concepts expressed in natural language) about the world are very much affected by their cognitive processes (e.g., Yokota, 2009), which is very natural considering that they are very cognitive products in humans. For example, people often perceive continuous forms among separately located objects so called spatial gestalts in the field of psychology and refer to them by such an expression as 'Five houses stand in line.' That is, people do not or cannot perceive the real world as it is. This fact means as well that knowledge representation without reflecting human cognition is not very useful for interaction between ordinary people and robots especially by natural language.

In the field of ontology, special attention has been paid to spatial language covering geography because its constituent concepts stand in highly complex relationships to underlying physical reality, accompanied with fundamental issues in terms of human cognition (for example, ambiguity, vagueness, temporality, identity, ...) appearing in varied subtle expressions (Harding, 2002). For the purpose of intuitive human-robot interaction, spatial language is also the most important of all sublanguages, especially when ordinary people and home robots must share knowledge of spatial arrangement of home utilities, such as desks, tables, etc.

Most conventional approaches to spatial language understanding have focused on computing purely objective geometric relations (i.e., topological, directional and metric relations) conceptualized as spatial prepositions or so, considering properties and functions of the objects involved (e.g., Egenhofer, 1991; Logan and Sadler, 1996; Shariff et al., 1998; Coventry et al., 2001). From the semantic viewpoint, spatial expressions have the virtue of relating in some way to visual scenes being described. Therefore, their semantic descriptions can be grounded in perceptual representations, possibly, cognitively inspired and coping with all kinds of spatial expressions including verb-centered ones such as S4-1 and S4-2 as well as preposition-centered ones such as S4-3.

(S4-1) The path *goes up* to the mountain top from our place.

(S4-2) The path *comes down* from the mountain top to our place.

(S4-3) The path is *between* us and the mountain top *over* us.

The verb-centered expressions, S4-1 and S4-2 are assumed to reflect not so much the purely objective geometrical relations but very much certain dynamism at human perception of the objects involved (Leisi, 1961; Langacker, 2005) because they can refer to the same scene in the external world referred to by the preposition-centered one, S4-3. This is also the case for S4-4 and S4-5 versus S4-6.

(S4-4) From the east, the road *enters* the city.

(S4-5) To the east, the road *exits* the city.

(S4-6) The west end of the road is *in* the city.

The fact stated above implies that conventional approaches to spatial language understanding will inevitably lead to a serious cognitive divide between humans and robots (in short, H-R cognitive divide) that causes miscommunication between them as shown in Fig. 4-4. In order to clear this hurdle, case-by-case treatments are not appropriate or efficient because there are a considerable number of such diversional uses of verb concepts. The author supposes that human active perception should affect the conceptualization of spatiotemporal events resulting in verb concepts such as *enter* and *exit*. Mental image directed semantic theory proposes its computational model in order for systematic removal of this kind

**Fig. 4-4.** Miscommunication due to H-R cognitive divide.

of H-R cognitive divide. This model is named *loci in attribute spaces*, to be introduced later, which has been intended to simulate mental image phenomena controlled by attention in human mind.

## 4.3 Semantics and Mental Image

Recently, deep neural networks have been achieving remarkable performance on various pattern-recognition tasks, especially visual classification problems. Simultaneously, interesting differences have been pointed out between human vision and current deep neural networks, raising questions about the generality of deep neural network computer vision because there exist images that are completely unrecognizable to humans, but that deep neural networks believe to be recognizable objects with 99.99% confidence (Nguyen et al., 2015). However, deep neural networks cannot accept any feedback from higher level cognition, such as reasoning, in order to calibrate misclassification because of the lack of any immediate means to convert knowledge representation (as awareness of misclassification by reasoning) into weight sets for connectionism. This may lead to such a claim that artificial intelligence should be more cognitive (Thompson, 2010; Kishimoto, 2016).

There are a considerable number of cognitively motivated studies on natural language semantics in association with mental image explicitly or implicitly (Leisi, 1961; Miller and Johnson-Laird, 1976; Sowa, 2000; Langacker, 2005). However, almost none of them are for natural language understanding because they lack systematic methodologies for both representation and computation of mental imagery. Many interesting researches on mental image itself in association with human thinking modes have been reported from various fields (Thomas, 2018) as well but none of them are from the viewpoint of natural language understanding. Quite distinguished from them, the author has proposed a mental image model in mental image directed semantic theory for systematic representation and computation of natural language semantics, more broadly, human knowledge. This section refers to several essential characteristics of mental images of the physical world as human cognitive products leading to the basis of mental image directed semantic theory.

Originally, our mental images of the external world are acquired through our sensory systems. Therefore, it is worth considering our perceptual processes. As well known, we do not perceive it as it is. That is, our perception does not begin with objective data gained through artificial sensors but with subjective sensation intrinsically (or subconsciously) articulated with contours of involved objects and gestalts among them. Here, this kind of articulation is called *intrinsic articulation*, attributed to our subconscious propensities toward the external world. Then, at the next stage, as active perception, we work our attention consciously to elaborate (or calibrate) intrinsic articulation by reasoning based on various kinds of knowledge and come to an interpretation of the sensation as a spatiotemporal relation among quasi-symbolic images (introduced in Chapter 1). This elaboration of intrinsic articulation is called *semantic articulation*.

As for intrinsic articulation, consider Fig. 4-5a and b. People sense continuous forms among separately located objects, so called spatial gestalts, and refer to them by such expressions as S4-7 and S4-8, respectively. Of course, we will become aware of the fact that these forms do not actually exist as soon as we focus our attention upon them. It is considered that the intrinsic articulation mechanism has been essential for the human species to survive since their primitive age. For example, the automatic or unconscious sensation of contours and gestalts can help people in their instant detection of dangerous edges of physical objects and a continuous path for escape from enemies appearing as discrete spots on the ground in such a scenario as Fig. 4-6. Of course, however, such intrinsic articulation always poses the risk of misrecognition as well. The neural network architectures prevailing today have the same risks because their simple-minded algorithms are essentially there to provide a machine with intrinsic articulation but not semantic articulation.

(S4-7) Five disks are in *line*.

(S4-8) Nine disks are placed in *the shape of X*.

As for semantic articulation, consider the expressions S4-9 and S4-10. People would intuitively and easily understand them so that they should describe the same scene in the external world as shown in Fig. 4-7. This is also the case for

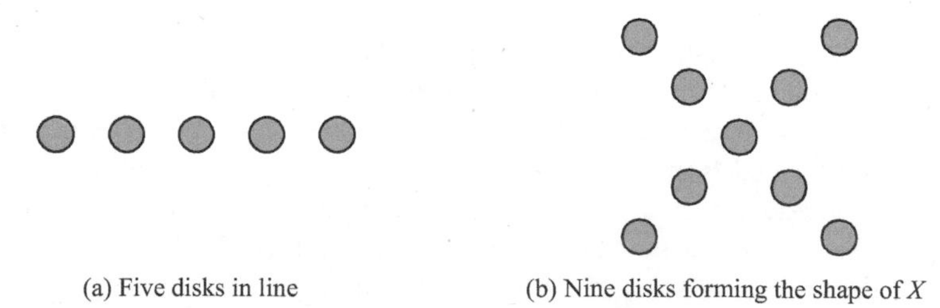

(a) Five disks in line (b) Nine disks forming the shape of *X*

**Fig. 4-5.** Gestalts sensed among multiple disks.

**Fig. 4-6.** Intrinsic articulation significant for survival.

S4-11 and S4-12. These pictures are actually the author's quasi-symbolic image representations while they are intended to look just like snapshots by a camera so as to be free of human dynamic cognitive processes.

(S4-9) The path *rises* from the stream to the hill.

(S4-10) The path *sinks* to the stream from the hill.

(S4-11) Route A and Route B *meet* at the city.

(S4-12) Route A and Route B *separate* at the city.

As known by these depicted scenes, it is apparent that the sentences do not reflect the purely objective geometrical relations very much. Instead, they seem to be subjective to human mental activity at cognition of the objects involved since their more neutral or objective descriptions would be like S4-13 and S4-14.

(S4-13) The stream is beneath the hill, and the path is between them.

(S4-14) Route A and Route B are contiguous at the city.

It is assumed that the verbs used in S4-9–S4-12 should mirror the movements of the focus of attention of the observer upon the objects involved in the external scenes. That is, what moves is the focus of the observer's gaze and each verb reflects the difference of its movement in direction, as shown in Fig. 4-8a and b, illustrating the focus of attention of the observers running along the path and the routes in the opposite directions (i.e., 'up and down', and 'left and right'), respectively.

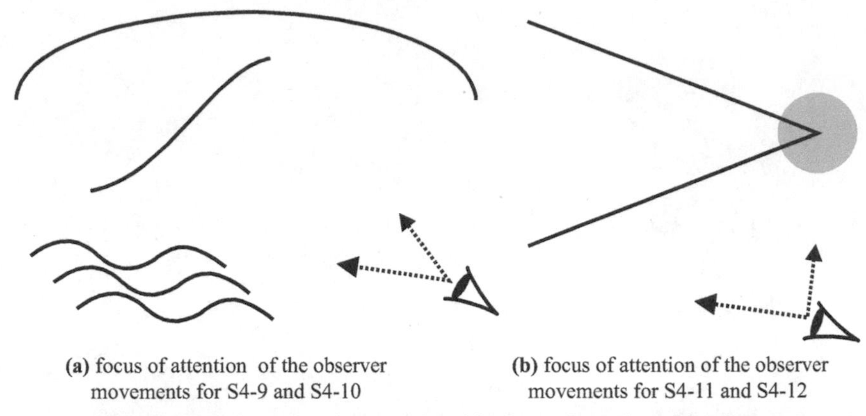

**(a)** focus of attention of the observer movements for S4-9 and S4-10

**(b)** focus of attention of the observer movements for S4-11 and S4-12

**Fig. 4-7.** Spatial scenes and movements of the focus of attention of the observer.

It seems extremely difficult for robots to reach such a paradoxical understanding in a systematic way. As mentioned above, spatial expressions have the virtue of relating in some way to visual scenes being described. Therefore, their semantic descriptions can be grounded in perceptual representation as mental image, possibly, cognitively inspired and coping with all kinds of spatial expressions, including such verb-centered ones as S4-9–S4-12, as well as preposition-centered ones. Actually, this is the author's motivation to develop mental image directed semantic theory but it is intended to be a general theory for systematic representation and computation of every kind of knowledge, including natural language semantics.

## 4.4 Quasi-Symbolic Images and Human Concept System

People understand or affect the world through its internal representation system, i.e., concept system so called here and depicted in Fig. 4-8, where natural semantics are the subsystem used to manage natural language meanings. It is supposed that the concept system consists of mental images of matters with verbal tokens or non-verbal tokens—quasi-symbolic images introduced in Chapter 1. According to our experiences, for example, the verbal token *dog* may remind us of a special dog such as *Snoopy*, but we know that it is a quasi-symbolic image, like a pictogram representing all dogs that we have ever encountered or not. This is because it is almost impossible for us to imagine or depict a perfectly general image of something, namely, *conceptual image*. Just try to imagine a *dog in general* and we will come to know that we can have at best a set of images of special dogs and tag it with the verbal token 'dog' to denote its actual meaning instead of conceptual image. This is also the case that someone's face image recalled by their (= his/her) name as a verbal token is a quasi-symbolic image, both of which are linked to all their information. Therefore, natural semantics in the concept system in Fig. 4-8 consists of both types of tokens and the rest, exclusively of quasi-

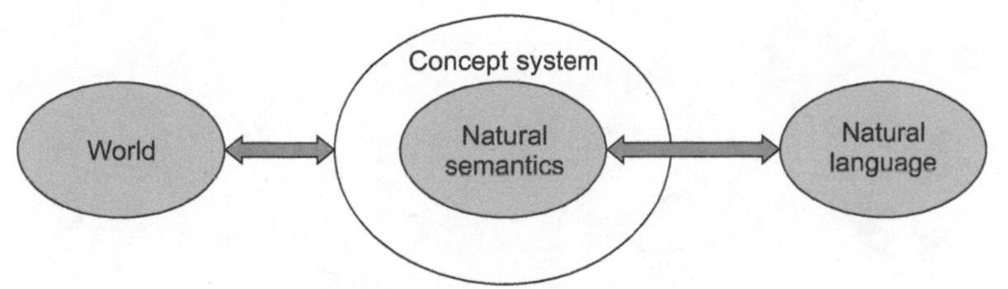


**Fig. 4-8.** Human concept system.

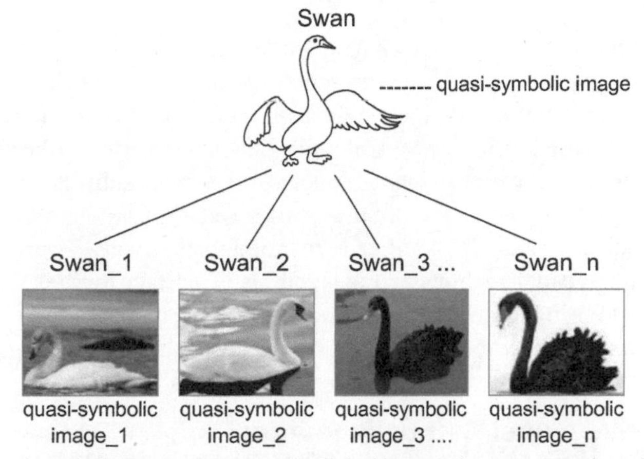


**Fig. 4-9.** Quasi-symbolic image of *swan* and its linked instance images as quasi-symbolic images.

Color version at the end of the book

symbolic images. Quasi-symbolic images are to work both as symbols and as virtual realities of matters while they are as general as concepts. For example, the word *swan* reminds the author of a *white* bird as a quasi-symbolic image, as shown in Fig. 4-9, and he affirms the assertion S4-15 consulting the quasi-symbolic image but it is not the case for S4-16 because he immediately remembers a counter instance image (quasi-symbolic image) of a *black* swan linked to the quasi-symbolic image of *swan*. Note that a quasi-symbolic image is semantically articulated well or precisely enough for us to affirm or deny such assertions as S4-15 and S4-16 as necessity although that is possible within the breadth and depth of our knowledge.

(S4-15) A swan has wings and webbed feet.

(S4-16) A swan is white.

The quasi-symbolic image of a matter is to be subjective to its holder and to be automatically or subconsciously generated from and assigned to a set of sensory images of the matter in their cognitive processes. It is thought that so-called 'tacit knowledge' is comprised of quasi-symbolic images and, thereby, the human

**Fig. 4-10.** Mental images of a *car* sketched by subjects of our psychological experiment.

species can perform subconscious knowledge computations (i.e., inferences) without conscious use of external representation means, such as language.

It is supposed that people must have a certain capability to manage both the types of quasi-symbolic images seamlessly. This hypothesis has been supported by our psychological experiments conducted on human subjects. For example, most of them had a quasi-symbolic image of a sedan or a coupe evoked by the word *car*, as shown Fig. 4-10. As already mentioned, these quasi-symbolic images worked both as virtual realities and as symbols. This fact implies that a certain automatic system to assign quasi-symbolic images to matters under attention or consideration should be implemented in a cognitive robot.

## 4.5 Primitive Quasi-Symbolic Images

A quasi-symbolic image that does not undergo semantic articulation any further is called primitive quasi-symbolic image. Primitive quasi-symbolic images are to correspond with live or recalled sensations, so called qualia, for example, 'redness' and 'sweetness'. According to the author's psychological experiences, the word 'red' reminds him of the representative quale, as shown in Fig. 4-11.

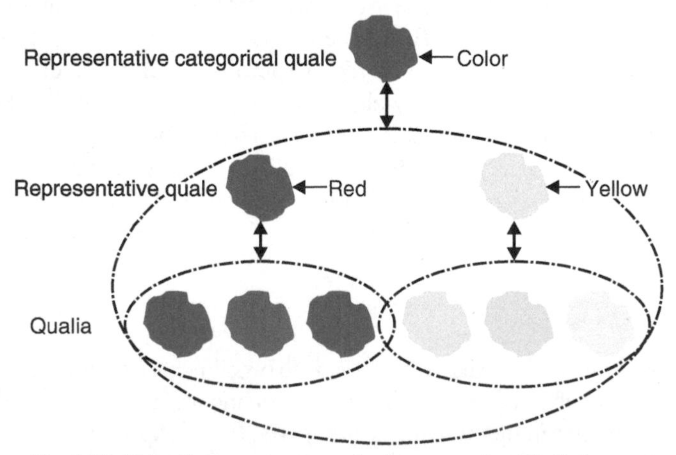


**Fig. 4-11.** Yokota's concept system of colors as quasi-symbolic images.

Color version at the end of the book

This fact implies that he has this quale as the primitive quasi-symbolic image of the word 'red'. Furthermore, the same quale is also his quasi-symbolic image of the category name, 'color'. The quasi-symbolic image (more exactly primitive quasi-symbolic image) of 'red' works well enough to understand S4-17 and such correctly. However, it is not the case for such an ordered pair of sentences as S4-18 and S4-19, where he is forced to replace the quasi-symbolic image of 'red', evoked by S4-18, with the quasi-symbolic image of 'blue' immediately after reading S4-19. He'd like the readers to try these examples. If they have no such concrete quale of 'color' as default at S4-18 and, therefore, do not experience such replacement at S4-19, their quasi-symbolic image must be highly abstracted or symbolized. That is, the author's case and the other could be formulated as (4-4) and (4-5), respectively, where *v* is a variable quale of color. The two cases can be explained consistently as reasoning based on such a knowledge piece as (4-6), where QSI is the abbreviation of quasi-symbolic image. That is, he has a special quasi-symbolic image, namely, a mental image of red, at the underlined part of (4-6) but the other does not possess any special quasi-symbolic image or the part at all.

(S4-17) Orange is similar to red more than green.

(S4-18) Tom is holding a color pencil.

(S4-19) It is for drawing in blue.

color(pencil, red) ==> color(pencil, blue). (4-4)

color(pencil, v) ==> color(pencil, blue). (4-5)

<u>color(pencil, QSI)</u>→ (∃v)color(pencil, v). (4-6)

## 4.6 Perception of Causality

We cannot sense abstract matters, such as kinetic force and electricity, existing remotely. For example, when we see such a surprising scene shown in Fig. 4-12,

**Fig. 4-12.** Which is carrying the other, the airplane or the person?

we do not believe that the man is pushing or carrying the airplane in the air but contrarily that it is flying by itself and that he is hanging onto it. That is, we believe that certain pulling force from the airplane is imposed upon the man. In this reasoning, we employ our commonsense knowledge of the ordinary world, that is, he is not Superman, while we cannot know the real kinetic relation between them. This example convinces us that our mental image can as well contain and indicate insensible *causal* relations resulted from our semantic articulation.

People can be aware of insensible or unobservable relations, such as kinetic causality, that can be conceptualized as a transitive verb. For example, consider the transitive verb '($x$) move ($y$)', referring to S4-20–S4-23. As easily imagined, S4-21–23 are what instantiate the vague or abstract scene of S4-20 at its kinetic effect from Tom upon the ball. This implies that kinetic effect must be symbolized for being ultimately abstract in the mental image.

(S4-20) Tom moved the ball.

(S4-21) Tom moved the ball <u>by throwing it</u>.

(S4-22) Tom moved the ball <u>by kicking it</u>.

(S4-23) Tom moved the ball <u>by rolling it</u>.

In turn, this fact gives a good reason for assigning a certain special symbol to an abstract matter in mental image. For example, mental image directed semantic theory depicts causal effect of $x$ upon $y$ as $x \rightarrow y$ in its graphical model of mental image called *Loci in Attribute Spaces*. The arrow is considered as one of primitive quasi-symbolic images representing an abstract relation. This necessarily implies that a primitive concept, such that '*x do something to y*', should be introduced to the formal system for representation and computation of mental image because it is unobservable and therefore can no longer be semantically articulated.

## 4.7 Semantic Articulation and Quasi-symbolic Image Connectors

As already mentioned, the author has such a hypothesis that our perception begins with sensation as a complex of spatial or temporal gestalts, namely, intrinsically articulated stimuli from the external world, and ends with a semantically articulated mental image (i.e., quasi-symbolic image) as an interpretation of the very sensation. At semantic articulation of sensation, we work our attention consciously in order to elaborate intrinsic articulation by reasoning, based on various kinds of knowledge, and come to an interpretation of the sensation as a spatiotemporal relation among quasi-symbolic images. Such elaboration is formulated as (4-7) where $\psi(X)$ is called 'semantic interpretation of sensation $X$'.

$$\psi(X)=R(x_1,\ldots,x_n), \tag{4-7}$$

where

$\psi$: Semantic articulation function provided in the human mind,

X: Sensation with intrinsic articulation,

$x_i$: An object as a quasi-symbolic image resulting from semantic articulation of X,

R(…): A total quasi-symbolic image for the spatiotemporal relation (e.g., between, move) among or the property (e.g., bitter, cold) of partial quasi-symbolic images such as objects (e.g., flower, ice) resulting from semantic articulation.

Sensation $X$ is just like a 3D movie with an absolute time interval as its duration and, therefore, the total interpretation of a set of sensations $\{X_1, X_2, \ldots, X_n\}$ ordered in time is a logical combination of their interpretations $\{\psi(X_1), \psi(X_2), \ldots, \psi(X_n)\}$ on the time axis, as formulated by (4-8), where $\wedge$ denotes logical AND (i.e., conjunction).

$$\psi(\{X_1, X_2, \ldots, X_n\}) = \psi(X_1) \wedge \psi(X_2) \wedge \ldots \wedge \psi(X_n) \quad (4\text{-}8)$$

In the author's consideration, the duration of a sensation is an absolute time interval, over which the observer's attention is put on the corresponding phenomenon outside the mind. Just a reflection will convince us that people cannot measure the absolute time by any chronograph but a certain relative time. Actually, they do not always consult a chronograph even if they can. Instead, they are aware of and memorize the temporal orders of events they experience and, therefore, mental image directed semantic theory has employed *tempo-logical connectives* as quasi-symbolic images to denote both logical and temporal relations between sensations. For example, imagine the scenario of S4-24 and we can easily come to know that S4-25 is its most plausible paraphrase among the three expressions, S4-25–S4-27. That is, the conjunction 'and' in S4-24 is not a purely logical AND but a tempo-logical AND as well as 'before', 'as' and 'after' involved here.

(S4-24) Tom washed his hands and took his lunch.

(S4-25) Tom washed his hands before he took his lunch.

(S4-26) Tom washed his hands as he took his lunch.

(S4-27) Tom washed his hands after he took his lunch.

## 4.8 Negation of Mental Image

Mental image of negated event of $A$ (i.e., $\sim A$) is not the image of a special event different from $A$ but the image of event $A$ itself with a tag implying 'negation' corresponding with the symbol '~'. On principle, there can be numerous candidates for $\sim A$ because they have only to be different from the image of $A$, for example, the image of 'walking' or 'swimming' for 'not running'. That is, negation of some event $A$ can be represented by a quasi-symbolic image which is a special example

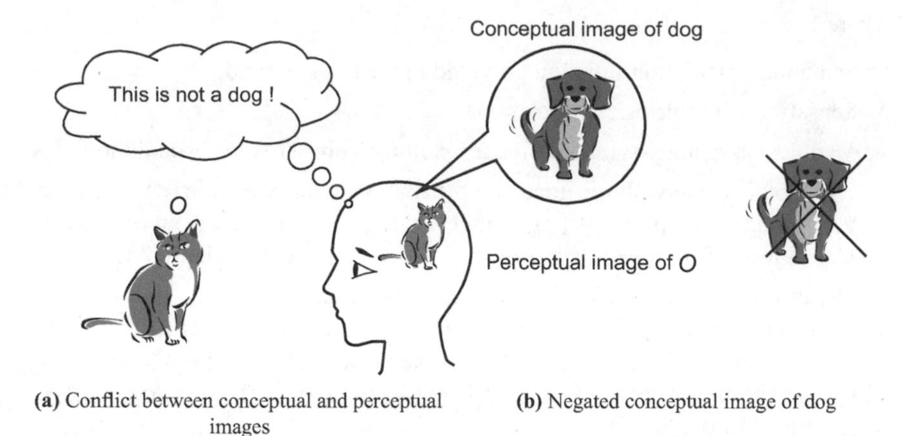


**(a)** Conflict between conceptual and perceptual images

**(b)** Negated conceptual image of dog

**Fig. 4-13.** Negation of mental image.

of ~*A*. However, we are quite aware that actually it represents ~*A*. According to our psychological experiments concerning negated mental image, the subjects drew a special example of ~*A* or a special symbol '×' that is conventional as the special token for negation in Japan. In the former case, however, the subjects involved were quite conscious about its representativeness for ~*A* as one of quasi-symbolic images and this consciousness corresponds with the symbol '~'.

Suppose such a situation as the observer uttering 'This is not a dog' at the scene shown in Fig. 4-13a. It is very trivial that he must compare the two images, namely, the perceptual image of object *O* and the conceptual image of dog. It is notable he could have uttered 'This is a cat'. However, the former utterance can be rather more natural than the latter if he is looking for a dog with its conceptual image as a quasi-symbolic image in mind. On the other hand, what mental images do we have evoked by this negative expression? Logically, we must have at least two images, namely, one quasi-symbolic image for dog and the other quasi-symbolic image for the referent of 'this'. In this case of our experiment, most of the subjects drew only one quasi-symbolic image for dog with a special symbol '×' overlapped as shown in Fig. 4-13b. This provides a good reason for employing this negation symbol as a primitive quasi-symbolic image in mental image directed semantic theory.

## 4.9 Imaginary Space Region

Mental image directed semantic theory is based on two major hypotheses about mental image. One is that mental image is in one-to-one correspondence with the focus of attention of the observer movement as mentioned above, and the other is that it is not one-to-one reflection of the real world. It is well-known that people perceive more than reality, for example, 'gestalt' so called in psychology. A psychological matter here is not a real matter but a product of human mental functions, including gestalt and abstract matters such as 'society', 'information', etc., in a broad sense. Especially, mental image directed semantic theory employs an imaginary object called imaginary space region in order for systematic formulation of mental image, perceptual or conceptual, of the physical world. For example, the linear region and the X-shaped region in Fig. 4-5 are imaginary space regions formed subconsciously within our perceptual image.

# 5 Computational Model of Mental Image

A considerable number of studies on mental image have been reported from various fields (Thomas, 2018). To the author's best knowledge, however, no model of mental image has been proposed from the viewpoint of computer science. His issues about mental image, presented in the previous chapters, can be summarized as the requirements for its computational model to satisfy M1–M4, as follows.

(M1) Intuitively acceptable and plausible in such a way as schematization.

(M2) Abstract and general enough to represent concepts of any kind, especially, 4D concepts and causality.

(M3) Recursive semantic articulation (with certain atomic images and their connective relations for infinite description) of mental image of any kind, based on a certain set of generative rules.

(M4) Simulation of human mental image operation and computation, including semantic re-articulation.

This chapter introduces *Loci in Attribute Spaces,* Yokota's original mental image model designed to satisfy these requirements.

## 5.1 Atomic Locus as Primitive Quasi-symbolic Image

Mental image directed semantic theory has proposed a mental image model as a quasi-symbolic image system to be formalized in predicate logic. In mental image directed semantic theory, word meanings are treated in association with mental images, not limited to visual but omnisensory, modeled as so-called *Loci in Attribute Spaces* which have extremely abstract quasi-symbolic images. See Fig. 5-1a and assume that the person is observing the phenomenon where the triangular gray object is moving in the sinusoidal trajectory and that its corresponding sensations (i.e., sensory images) are being caused in the observer's mind. In this case, the moving triangular gray object is assumed to be perceived as the loci in the three attribute spaces, namely, those of *Location*, *Color* and *Shape* in his mind. As easily imagined, attribute spaces correspond with human sensory systems and the loci represent certain sensations of the phenomena outside or

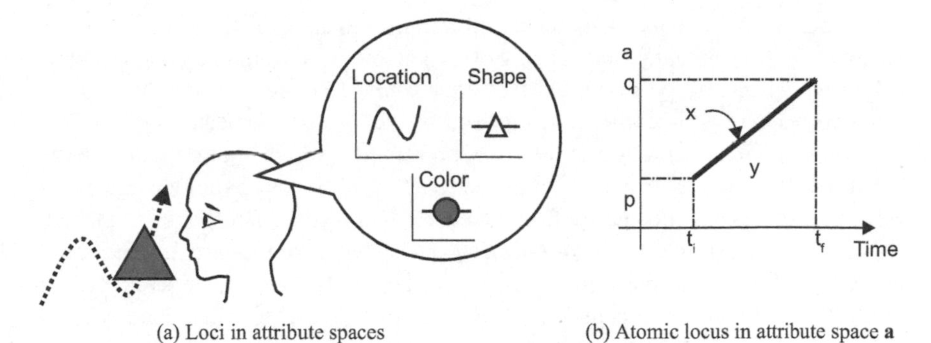


(a) Loci in attribute spaces (b) Atomic locus in attribute space **a**

**Fig. 5-1.** Mental image model.

inside human minds. From the viewpoint of artifact, an attribute space stands for a certain measuring instrument or sensor, like a chronograph, barometer or thermometer, and the loci represent the movements of its indicator.

These loci are to be articulated by *Atomic Locus* as a primitive quasi-symbolic image over a certain *absolute time* interval $[t_i, t_f]$, as depicted in Fig. 5-1b, and formulated as (5-1) in the knowledge representation language $\boldsymbol{L}_{md}$ (i.e., mental image description language), where the interval is suppressed because people cannot know absolute time, only a certain *relative time* indicated by a chronograph. Mental image description language (i.e., $\boldsymbol{L}_{md}$) is to be rigidly introduced in the next chapter.

$$L(x,y,p,q,a,g,k) \quad (5\text{-}1)$$

The expression (1) works as a formula in many-sorted predicate logic, where *L* is a predicate constant with five types of terms: *Matter* (at *x* and *y*), *Value* (at *p* and *q*), *Attribute* (at *a*), *Pattern* (*or Evet Type*) (at *g*) and *Standard* (at *k*). Conventionally, Matter variables are headed by *x*, *y* and *z*. This formula is called *Atomic Locus Formula* and its first two arguments are sometimes referred to as Event Causer and Attribute Carrier, respectively, while event causers are often optional in natural concepts such as intransitive verbs. For simplicity, the syntax of $\boldsymbol{L}_{md}$ allows Matter terms (e.g., *Tokyo* and *Osaka* in (5-2) and (5-3)) to appear at Values or Standard in order to represent their values in each place at the time or over the time-interval. Moreover, when it is not so significant to discern event causers or Standards, they are often denoted by anonymous variables, usually symbolized as '_' for the sake of simplicity. A logical combination of atomic locus formulas defined as a well-formed formula (i.e., wff) in predicate logic is called simply *locus formula*.
The intuitive interpretation of (5-1) is given as follows.

***Matter 'x' causes Attribute 'a' of Matter 'y' to keep* (p=q) *or change* (p ≠ q) *its values temporally* (g=Gt) *or spatially* (g=Gs) *over a certain absolute time-interval, where the values 'p' and 'q' are relative to the standard 'k'.***

When $g=G_t$, the locus indicates monotonic change (or constancy) of the attribute in time domain, and when $g=G_s$, that in space domain. The former is called *temporal change event* and the latter, *spatial change event*. These event types correspond with *temporal gestalt* and *spatial gestalt*, respectively.

For example, the motion of the *bus* represented by S5-1 is a temporal change event and the ranging or extension of the *road* by S5-2 is a spatial change event whose meanings or concepts are formulated as (5-2) and (5-3), respectively, where $A_{12}$ denotes the attribute *Physical Location*. These two formulas are different only at the term *Pattern*. In these formulas, the variable $x$ bound by ∃ at event causer implies 'there is an event causer but it is unknown', which is often denoted by a special symbol '_' placed there as defined rigidly later.

(S5-1) The bus runs from Tokyo to Osaka.

(S5-2) The road runs from Tokyo to Osaka.

$$(\exists x,y,k)L(x,y,Tokyo,Osaka,A_{12},G_t,k)\wedge bus(y) \quad (5\text{-}2)$$

$$(\exists x,y,k)L(x,y,Tokyo,Osaka,A_{12},G_s,k)\wedge road(y) \quad (5\text{-}3)$$

It has been often argued that human active sensing processes may affect perception and, in turn, conceptualization and recognition of the physical world (Leisi, 1961; Noton, 1970; Miller, 1976; Langacker, 2005). The author has hypothesized that the difference between temporal and spatial change event concepts can be attributed to the relationship between the attribute carrier and the focus of attention of the observer (Yokota, 2005). To be brief, it is hypothesized that the focus of attention of the observer is fixed on the attribute carrier as a point (i.e., without spatial extension) in a temporal change event but *runs* about on the attribute carrier in a spatial change event. Consequently, as shown in Fig. 5-2, the *bus* and the focus of attention of the observer move together in the case of S5-1, while the focus of the attention of the observer solely moves along the *road* in the case of S5-2. That is, ***all loci in attribute spaces are assumed to correspond one to one with movements or, more generally, temporal change events of the focus of attention of the observer***.

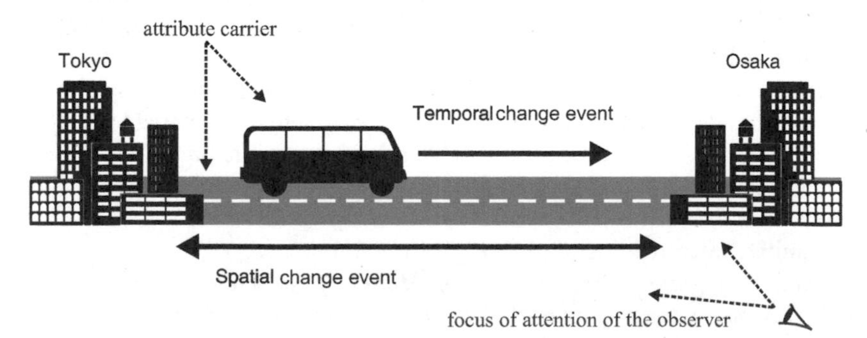


**Fig. 5-2.** Movements of the focus of attention of the observer and event types.

## 5.2 Temporal Conjunctions as Quasi-symbolic Image Connectors

The duration of a locus corresponds to an absolute time-interval, over which the focus of attention of the observer is put on the corresponding phenomenon outside or inside our mind. Such an absolute time-interval is suppressed in an atomic locus formula because it is assumed that people cannot measure the absolute time by any chronograph but a certain relative time. Actually, people do not always consult a chronograph even if they can. Instead, they are aware of and memorize the temporal sequences of events they experience. Mental image directed semantic theory has employed a set of *temporal conjunctions* denoting both logical conjunction (i.e., AND) and temporal relations between loci by themselves, because these must be considered simultaneously in locus articulation.

Temporal conjunction $\wedge_i$ is defined by (5-4), where $\tau_i$, $\chi$ and $\wedge$ represent one of the temporal relations indexed by an integer *i*, a locus, and an ordinary binary logical AND, respectively. The definition of $\tau_i$ is given in Table 5-1, from which the theorem (5-5) can be deduced. This table shows the complete list of topological relations between two intervals, where 13 types of relations are discriminated by $\tau_i$ $(-6 \leq i \leq 6)$. This is in accordance with Allen's notation (Allen, 1984), which is not for pure temporal relations but *temporal conjunctions* $(=\wedge_i)$, because each relation holds if and only if its two arguments are both true, namely, the two events actually happen.

**Table 5-1.** List of temporal relations.

<table>
<tr><th colspan="3">Temporal relations and definition of $\tau_i$[†]</th><th>Allen's notation</th></tr>
<tr><td rowspan="2">$\chi_1$ |··············|<br>$\chi_2$ |··············|</td><td rowspan="2">$t_{11}=t_{21} \wedge t_{12}=t_{22}$</td><td>$\tau_0(\chi_1, \chi_2)$</td><td>equals$(\chi_1,\chi_2)$</td></tr>
<tr><td>$\tau_0(\chi_2, \chi_1)$</td><td>equals$(\chi_2, \chi_1)$</td></tr>
<tr><td rowspan="2">$\chi_1$ |·······|<br>$\chi_2$         |·······|</td><td rowspan="2">$t_{12}=t_{21}$</td><td>$\tau_1(\chi_1, \chi_2)$</td><td>meets$(\chi_1, \chi_2)$</td></tr>
<tr><td>$\tau_{-1}(\chi_2, \chi_1)$</td><td>met-by$(\chi_2, \chi_1)$</td></tr>
<tr><td rowspan="2">$\chi_1$ |·······|<br>$\chi_2$ |··············|</td><td rowspan="2">$t_{11}=t_{21} \wedge t_{12}<t_{22}$</td><td>$\tau_2(\chi_1, \chi_2)$</td><td>starts$(\chi_1, \chi_2)$</td></tr>
<tr><td>$\tau_{-2}(\chi_2, \chi_1)$</td><td>started-by$(\chi_2, \chi_1)$</td></tr>
<tr><td rowspan="2">$\chi_1$    |·······|<br>$\chi_2$ |··············|</td><td rowspan="2">$t_{11}>t_{21} \wedge t_{12}<t_{22}$</td><td>$\tau_3(\chi_1, \chi_2)$</td><td>during$(\chi_1, \chi_2)$</td></tr>
<tr><td>$\tau_{-3}(\chi_2, \chi_1)$</td><td>contains$(\chi_2, \chi_1)$</td></tr>
<tr><td rowspan="2">$\chi_1$      |········|<br>$\chi_2$ |··············|</td><td rowspan="2">$t_{11}>t_{21} \wedge t_{12}=t_{22}$</td><td>$\tau_4(\chi_1, \chi_2)$</td><td>finishes$(\chi_1, \chi_2)$</td></tr>
<tr><td>$\tau_{-4}(\chi_2, \chi_1)$</td><td>finished-by$(\chi_2, \chi_1)$</td></tr>
<tr><td rowspan="2">$\chi_1$ |·······|<br>$\chi_2$            |·······|</td><td rowspan="2">$t_{12}<t_{21}$</td><td>$\tau_5(\chi_1, \chi_2)$</td><td>before$(\chi_1, \chi_2)$</td></tr>
<tr><td>$\tau_{-5}(\chi_2, \chi_1)$</td><td>after$(\chi_2, \chi_1)$</td></tr>
<tr><td rowspan="2">$\chi_1$ |·········|<br>$\chi_2$        |········|</td><td rowspan="2">$t_{11}<t_{21} \wedge t_{21}<t_{12} \wedge t_{12}<t_{22}$</td><td>$\tau_6(\chi_1, \chi_2)$</td><td>overlaps$(\chi_1, \chi_2)$</td></tr>
<tr><td>$\tau_{-6}(\chi_2, \chi_1)$</td><td>overlapped-by$(\chi_2, \chi_1)$</td></tr>
</table>

[†]The durations of $\chi_1$ and $\chi_2$ are $[t_{11}, t_{12}]$ and $[t_{21}, t_{22}]$, respectively.

$$\chi_1 \wedge_i \chi_2 \Leftrightarrow (\chi_1 \wedge \chi_2) \wedge \tau_i(\chi_1, \chi_2) \quad (5\text{-}4)$$

$$\tau_{-i}(\chi_2, \chi_1) \equiv \tau_i(\chi_1, \chi_2) \quad (\forall i \in \{0, \pm1, \pm2, \pm3, \pm4, \pm5, \pm6\}) \quad (5\text{-}5)$$

The temporal conjunctions are to be redefined as part of tempo-logical connectives again in the next chapter. Among them, SAND ($\wedge_0$) and CAND ($\wedge_1$) are used most frequently, standing for *Simultaneous AND* and *Consecutive AND*, conventionally symbolized as Π and •, respectively.

Consider Fig. 5-3, for example. Tom is observing the green truck to recognize that it is carrying the pyramidal rock from Tokyo to Osaka via Nagoya. His recognition here can be depicted as loci in Fig. 5-4a–c and formulated as (5-6), where $K_{Tom}$ denotes Tom's individual standard value and the symbol '_' stands for abbreviation of an unknown but existing value, while the concerned attributes are Physical location ($A_{12}$), Color ($A_{32}$), and Shape ($A_{11}$).

$$\begin{aligned}&(L(Truck,Truck,Tokyo,Nagoya,A_{12},G_t,K_{Tom})\Pi L(Truck,Rock,Tokyo,Nagoya,\\&A_{12},G_t,K_{Tom}))\bullet\\&(L(Truck,Truck,Nagoya,Osaka,A_{12},G_t,K_{Tom})\Pi L(Truck,Rock,Nagoya,Osaka,\\&A_{12},G_t,K_{Tom}))\Pi\\&((L(\_,Truck,Green,Green,A_{32},G_t,K_{Tom})\Pi L(\_,Rock,Pyramidal,Pyramidal,\\&A_{11},G_t,K_{Tom}))\end{aligned} \quad (5\text{-}6)$$

As easily understood, the properties of a tempo-logical connective depend on those of the purely logical connective and the temporal relations ($\tau_i$) involved. There are a considerable number of trivial theorems concerning temporal relations such as (5-7)–(5-14) below.

$$(\forall i)\tau_i(\chi_1,\chi_2)\wedge\tau_0(\chi_2,\chi_3) .\supset. \tau_i(\chi_1,\chi_3) \quad (5\text{-}7)$$

$$\tau_1(\chi_1,\chi_2)\wedge\tau_1(\chi_2,\chi_3) .\supset. \tau_5(\chi_1,\chi_3) \quad (5\text{-}8)$$

$$\tau_1(\chi_1,\chi_2)\wedge\tau_3(\chi_2,\chi_3) .\supset. \tau_5(\chi_1,\chi_3) \quad (5\text{-}9)$$

$$\tau_1(\chi_1,\chi_2)\wedge\tau_4(\chi_2,\chi_3) .\supset. \tau_5(\chi_1,\chi_3) \quad (5\text{-}10)$$

$$\tau_1(\chi_1,\chi_2)\wedge\tau_5(\chi_2,\chi_3) .\supset. \tau_5(\chi_1,\chi_3) \quad (5\text{-}11)$$

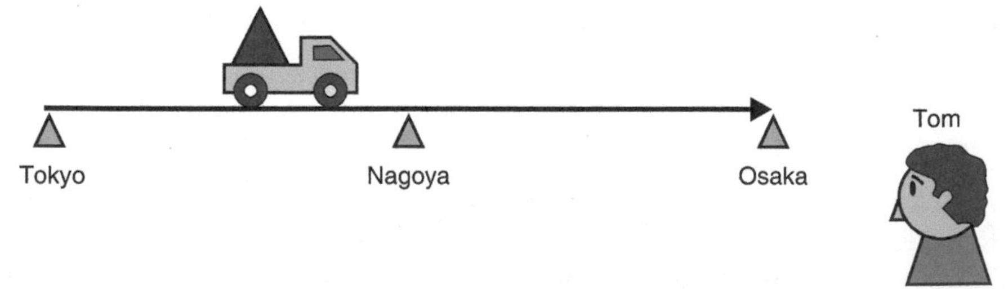


**Fig. 5-3.** A scene and the observer, Tom.

Color version at the end of the book

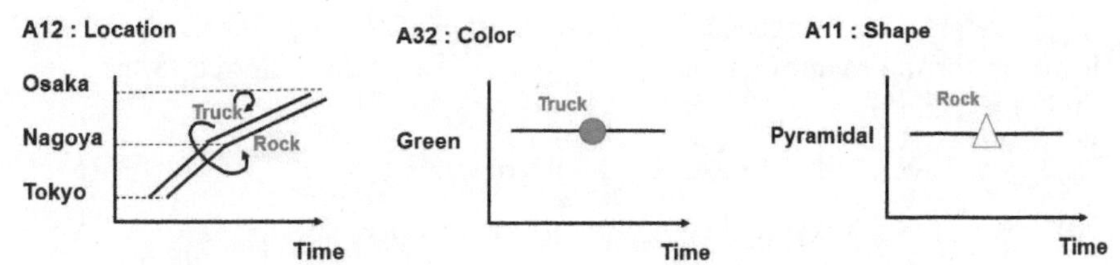


**Fig. 5-4.** Loci in attribute spaces as Tom's perception of the scene at Fig. 5-3.

$$\tau_1(\chi_1,\chi_2)\wedge\tau_6(\chi_2,\chi_3) .\supset. \tau_5(\chi_1,\chi_3) \quad (5\text{-}12)$$

$$\tau_2(\chi_1,\chi_2)\wedge\tau_1(\chi_2,\chi_3) .\supset. \tau_5(\chi_1,\chi_3) \quad (5\text{-}13)$$

$$\tau_5(\chi_1,\chi_2)\wedge\tau_1(\chi_2,\chi_3) .\supset. \tau_5(\chi_1,\chi_3) \quad (5\text{-}14)$$

As easily calculated, $\tau_k(\chi_1,\chi_3)$ in (5-15) is not always determined uniquely.

$$\tau_i(\chi_1,\chi_2)\wedge\tau_j(\chi_2,\chi_3) .\supset. \tau_k(\chi_1,\chi_3) \quad (5\text{-}15)$$

## 5.3 Empty Event

We would often say "Nothing happened" but at least time passed. Then, in order for the explicit indication of time points, a very important concept called *Empty Event*, denoted by ε is, introduced. An empty event stands only for absolute time elapsing and is explicitly defined as (5-16) with the attribute *Time Point* ($A_{34}$) and the standard $S_{Ta}$ denoting absolute time. This formula reads that time passes absolutely by itself, which is the author's intuition but can only be assumed. According to this scheme, the absolute time duration $[t_a, t_b]$ of an arbitrary locus χ can be expressed as (5-17) or (5-17'). However, actually, a certain relative time is to be employed as the movement (i.e., locus) of the indicator of a chronograph where another standard value is to be put instead of $S_{Ta}$ in (5-16) because of the impossibility of measuring absolute time. That is, for example, Greenwich Mean Time of event χ is to be indicated as (5-18) with a relative time duration $[t_c, t_d]$.

$$\varepsilon([t_i,t_j]) \Leftrightarrow L(\text{Time},\text{Time},t_i,t_j,A_{34},G_t,S_{Ta}), \quad (5\text{-}16)$$

where $[t_i,t_j] \in \Delta$ (=$\{[p,q] \mid p \leq q$, p and q are real numbers$\}$).

$$\chi\ \Pi\ \varepsilon([t_a,t_b]). \quad (5\text{-}17)$$

$$\chi\ \Pi\ \varepsilon(\mathbf{d}), \text{ where } \mathbf{d} \in \Delta. \quad (5\text{-}17')$$

$$\chi\ \Pi\ L(\text{Time},\text{Time},t_c,t_d,A_{34},G_t,\text{GMT}). \quad (5\text{-}18)$$

By the way, such an intuitive expression as S5-3 concerns so-called 'psychological time' which is also discerned from certain objective time, such as Greenwich Mean Time, by using the standard parameter.

(S5-3) Time passes very quickly.

Therefore, more rigidly, (5-2) and (5-3) must be (5-2') and (5-3'), respectively. However, for the sake of simplicity, the former are often preferred to the latter without any confusion.

$$(\exists \mathbf{d})\ (\exists x,y,k)L(x,y,\text{Tokyo},\text{Osaka},A_{12},G_t,k)\Pi\varepsilon(\mathbf{d})\wedge \text{bus}(y) \qquad (5\text{-}2')$$

$$(\exists \mathbf{d})\ (\exists x,y,k)L(x,y,\text{Tokyo},\text{Osaka},A_{12},G_s,k)\Pi\varepsilon(\mathbf{d})\wedge \text{road}(y) \qquad (5\text{-}3')$$

Any pair of loci temporally related in certain attribute spaces can be formulated as (5-19)–(5-23) in exclusive use of SANDs, CANDs and empty events. For example, the loci shown in Fig. 5-5a and b correspond to the formulas (5-20) and (5-23), respectively. As easily understood, the connectivity law holds for each of SAND and CAND.

$$\chi_1 \wedge_2 \chi_2 \ .\equiv.\ (\chi_1 \bullet \varepsilon)\Pi\chi_2 \qquad (5\text{-}19)$$

$$\chi_1 \wedge_3 \chi_2 \ .\equiv.\ (\varepsilon_1 \bullet \chi_1 \bullet \varepsilon_2)\Pi\chi_2 \qquad (5\text{-}20)$$

$$\chi_1 \wedge_4 \chi_2 \ .\equiv.\ (\varepsilon \bullet \chi_1)\Pi\chi_2 \qquad (5\text{-}21)$$

$$\chi_1 \wedge_5 \chi_2 \ .\equiv.\ \chi_1 \bullet \varepsilon \bullet \chi_2 \qquad (5\text{-}22)$$

$$\chi_1 \wedge_6 \chi_2 \ .\equiv.\ (\chi_1 \bullet \varepsilon_3)\Pi(\varepsilon_1 \bullet \chi_2)\Pi(\varepsilon_1 \bullet \varepsilon_2 \bullet \varepsilon_3) \qquad (5\text{-}23)$$

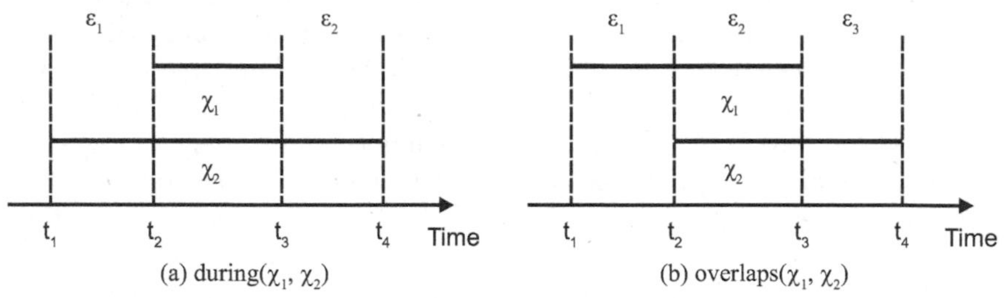


(a) during($\chi_1$, $\chi_2$) (b) overlaps($\chi_1$, $\chi_2$)

**Fig. 5-5.** Tempo-logical relations: *during* and *overlaps*.

## 5.4 Attributes and Standards

The attribute spaces for humans correspond to the sensory receptive fields in their brains. At present, about 50 attributes concerning the physical world have been extracted exclusively from English and Japanese words, as shown in Table 5-2. They are associated with all of the 5 senses (i.e., sight, hearing, smell, taste and touch) in our everyday life, while those for information media other than languages correspond to limited senses. For example, those for pictorial media, marked with '*' in Table 5-2, associate limitedly with the sense 'sight' as a matter of course. The attributes of this sense occupy the greater part of all, which implies that sight is essential for humans to conceptualize the external world. And this kind of classification of attributes plays a very important role in our cross-media operating system (Yokota, 2005). Viewed from rigidly scientific semantics,

**Table 5-2.** List of attributes.

| Code | Attribute [Property]† | Linguistic expressions for attribute values |
|---|---|---|
| A. Attributes concerning the physical world. | | |
| *A01 | WORLD [N] | Many whales were flying in my dream. |
| *A02 | LENGTH [S] | The stick is 2 meters long (in length). |
| *A03 | HEIGHT [S] | The tree is 2 meters high (in height). |
| *A04 | WIDTH [S] | The door is 2 meters wide (in width). |
| *A05 | THICKNESS [S] | The book is 2 centimeters thick (in thickness). |
| *A06 | DEPTH1 [S] | The swimming pool is 2 meters deep (in depth). |
| *A07 | DEPTH2 [S] | The cave is 100 meters deep (in depth). |
| *A08 | DIAMETER [S] | The hole is 2 meters across (in diameter). |
| *A09 | AREA [S] | The crop field is 10 square miles. |
| *A10 | VOLUME [S] | The box is 10 cubic meters. |
| *A11 | SHAPE [N] | The cake is round. |
| *A12 | PHYSICAL LOCATION [N] | Tom moved to Tokyo. |
| *A13 | DIRECTION [N] | The box is to the left of the chair. |
| *A14 | ORIENTATION [N] | The door faces to south. |
| *A15 | TRAJECTORY [N] | The plane circled in the sky. |
| *A16 | VELOCITY [S] | The boy runs very fast. |
| *A17 | MILEAGE [S] | The car ran ten miles. |
| A18 | STRENGTH OF EFFECT [S] | He is very strong. |
| A19 | DIRECTION OF EFFECT [N] | He pulled the door. |
| A20 | DENSITY [S] | The milk is very dense. |
| A21 | HARDNESS [S] | The candy is very soft. |
| A22 | ELASTICITY [S] | The sheet is elastic. |
| A23 | TOUGHNESS [S] | The glass is very brittle. |
| A24 | FEELING [S] | The cloth is smooth. |
| A25 | MOISTURE [S] | The paint is still wet. |
| A26 | VISCOSITY [S] | The liquid is oily. |
| A27 | WEIGHT [S] | The metal is very light. |
| A28 | TEMPERATURE [S] | It is hot today. |
| A29 | TASTE [N] | The grapes here are very sour. |
| A30 | ODOUR [N] | The gas is pungent. |
| A31 | SOUND [N] | His voice is very loud. |
| *A32 | COLOR [N] | The apple is red. Tom painted the desk white. |
| A33 | INTERNAL SENSATION [N] | I am very tired. |

*Table 5-2 contd. ...*

... *Table 5-2 contd.*

| **Code** | **Attribute [Property]†** | **Linguistic expressions for attribute values.** |
|---|---|---|
| A34 | TIME POINT [S] | It is ten o'clock. |
| A35 | DURATION [S] | He studies for two hours every day. |
| A36 | QUANTITY [S] | Here are many people. |
| A37 | ORDER [S] | Tom sat next to Mary. |
| A38 | FREQUENCY [S] | He did it twice. |
| A39 | VITALITY [S] | The old man still alive. |
| A40 | SEX [S] | The operator is female. |
| A41 | QUALITY [N] | We make cheese from milk. |
| A42 | NAME [N] | The father named his baby Thomas. |
| A43 | CONCEPTUAL CATEGORY [N] | A whale is a mammal. |
| *A44 | TOPOLOGY [N] | Tom went out of the room. |
| *A45 | ANGULARITY [S] | The knife is dull. |
| B. Attributes concerning the mental world. | | |
| B01 | WORTH [N] | The house was damaged (improved, broken, etc.). |
| B02 | LOCATION OF INFORMATION [N] | We think (tell, read, etc.), that .... |
| B03 | EMOTION [N] | I like (love, respect, etc.), him. |
| B04 | BELIEF VALUE [S] | I believe (wish, decide, etc.), that .... |
| B05 | TRUTH VALUE [S] | I know (realize, understand, etc.), that ... |
| B06 | LOCATION OF OWNERSHIP [N] | I gave him a book. I sold a book to him. |
| B07 | LOCATION OF USERSHIP [N] | I lent him a book. I borrowed a book from him. |

†S: scalar attribute, N: non-scalar attribute. *Attributes concerning the sense of sight.

**Table 5-3.** List of standards.

| **Categories of standards** | **Remarks** |
|---|---|
| Rigid Standard | Objective standards such as denoted by measuring *units* (meter, gram, etc.). |
| Species Standard | The *attribute value ordinary* for a species. A *short train* is ordinarily longer than a *long pencil.* |
| Proportional Standard | '*Oblong*' means that the width is greater than the height at a physical object. |
| Individual Standard | *Much* money for one person can be too *little* for another. |
| Purposive Standard | One room large enough for a person's *sleeping* might be too small for his *jogging*. |
| Declarative Standard | Gives the origin of deviation in an attribute. Tom is taller *than Jim*. The origin of an order such as 'next' must be declared explicitly, as in 'next *to him*'. |
| Tacit Standard | Gives granularities for semantic articulation of loci. Most of them are tacit due to non-linguistic cognitive processes working as the units of cognitive scales. |

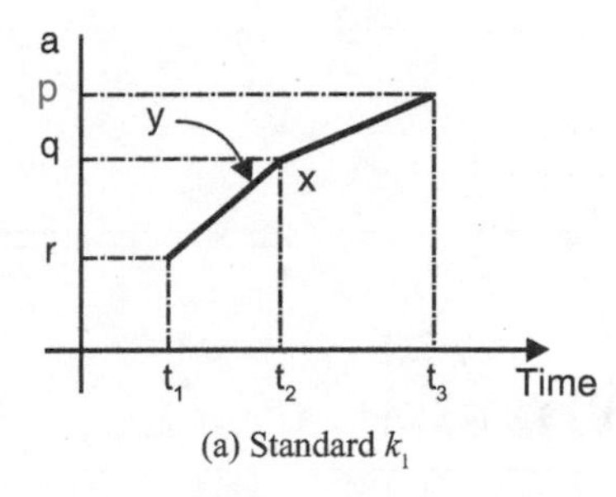


(a) Standard $k_1$

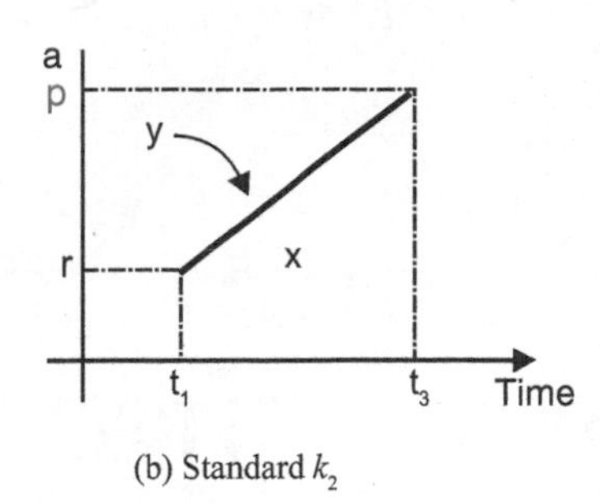


(b) Standard $k_2$

**Fig. 5-6.** Arbitrariness of locus articulation due to standards: Standard $k_1$ (a) is finer than $k_2$ (b).

however, some part of these attributes are not independent, which should be improved or treated carefully at computation.

Correspondingly, seven categories of standards shown in Table 5-3 have been extracted that are assumed necessary for representing values of each attribute in Table 5-2, encompassing the conventional categorization (Leisi, 1961). In general, the attribute values represented by words are relative to certain standards, as explained briefly in Table 5-3. For example (5-24) and (5-25), are different formulations of a locus due to the different Tacit Standards $k_1$ and $k_2$ for scaling, as shown in Fig. 5-6a and b, respectively. That is, whether the point $(t_2, q)$ is significant or not, more generally, how to articulate a locus depends on the precisions or the granularities of these standards.

$$(L(y,x,p,q,a,g,k_1)\Pi\ \varepsilon([t_1,t_2]))\bullet(L(y,x,q,r,a,g,k_1)\Pi\ \varepsilon([t_2,t_3])) \quad (5\text{-}24)$$

$$L(y,x,p,r,a,g,k_2)\Pi\ \varepsilon([t_1,t_3]) \quad (5\text{-}25)$$

# 6 Formal System

This chapter presents a formal system for representation and computation of human common-sense knowledge about the physical world to be employed in robotic natural language understanding. A formal system is defined as a pair of a formal language and a deductive system consisting of the axioms and inference rules employed for theorem derivation. Mental image description language (i.e., $\boldsymbol{L}_{md}$) is a formal language for many-sorted predicate logic with five types of terms specific to the mental image model. Therefore, the deductive system intended here is to be based on the deductive apparatus for predicate logic.

## 6.1 Semantic Principle of Mental Image Description Language

Mental image directed semantic theory is based on the presumption that all pieces of human knowledge of the physical world are to be reduced to attribute values of matters and their spatiotemporal relations, namely, loci in attribute spaces as mental images. An attribute space is assumed to be one kind of tolerance space (Zeeman, 1962) on which sensations are to be mapped with a distance in proportion to their similarity. A tolerance $\tau$ on a set $O$ is a relation $\tau$ that is reflexive and symmetric. A set $O$ together with a tolerance $\tau$ is called a tolerance space (denoted $(O, \tau)$). An attribute value term is to refer to a region of the attribute space corresponding to a set of similar sensations. Therefore, a sensory word such as 'red' corresponds to a set of sensations with a certain degree of similarity called 'tolerance' which is located nearer to the 'orange' region than to the 'blue' region in the attribute space of Color ($A_{32}$). An atomic locus $L(x,y,p,q,a,g,k)$ is to represent a bundle of monotonic paths between regions $p$ and $q$ in the attribute space denoted by $(a,g,k)$, where the term 'monotonic' is not used mathematically but in an intuitive sense as being without significant regions on the way.

The formal language $\boldsymbol{L}_{md}$ is for integrative multimedia understanding, including robotic sensation and action, where the semantic primitives, namely, primitive quasi-symbolic images, are grounded in human sensory imagery modeled as loci in attribute spaces and, moreover, they are applied as constants in the deductive system for systematic computation. It is essential for robots to be provided with natural language semantics grounded in the world where they should sense or act. The conceptual or semantic primitives ever proposed for semantic representation and computation of natural language were for such natural language processing systems as automatic paraphrasers and translators, in that both input and output were texts (e.g., Schank, 1969; Wilks, 1972) and, therefore, they did not need to be grounded in the referent worlds.

## 6.2 Syntax of *Mental Image Description Language*

The symbols of $L_{md}$ for the deductive system are listed as (i)–(xiii) below. These symbols are possibly subscripted just like $A_{01}$, $G_s$, etc.

i) logical connectives: $\sim, \wedge, \vee, \supset, \equiv$
ii) quantifiers: $\forall, \exists$
iii) auxiliary constants: ., (, )
iv) sentence variables: $\chi$
v) individual variables
   a) matter variables: x, y, z
   b) attribute variables: a
   c) value variables: p, q, r, s, t
   d) pattern variables: g
   e) standard variables: k
vi) sentence constants: N
vii) predicate constants: L, $\tau$, =, $\neq$, >, < , and others to be introduced where needed
viii) individual constants
   a) matter constants: To be introduced where needed
   b) attribute constants: A, B
   c) value constants: To be introduced where needed
   d) pattern constants: G
   e) standard constants: K
xi) function constants: Arithmetic operators such as +, –, etc., and others to be introduced where needed
xii) meta-symbols: $\Leftrightarrow, \rightarrow, \leftrightarrow, \vdash$ and others to be introduced where needed
xiii) others: To be defined by the symbols above

Mental image description language employs one special predicate constant *L* called *atomic locus*, as already described in Chapter 5. The predicate *L* is such a seven-place predicate as is given by expression (6-1), called *atomic locus formula*, whose intuitive interpretation is given in the same way as presented in the previous chapter.

$$L(\omega_1,\omega_2,\omega_3,\omega_4,\omega_5,\omega_6,\omega_7) \quad (6\text{-}1)$$

The formula (6-1) is a well-formed *atomic locus formula* if and only if the conditions (a)–(g) are satisfied. A logical combination of atomic locus formulas is simply called *locus formula*. Except for this point, the syntactic rules are the same as those of the conventional predicate logic.

a) $\omega_1$ is a matter term (variable or constant)
b) $\omega_2$ is a matter term
c) $\omega_3$ is a value or a matter term
d) $\omega_4$ is a value or a matter term
e) $\omega_5$ is an attribute term
f) $\omega_6$ is a pattern term
g) $\omega_7$ is a standard (or matter) term

The intuitive interpretation of (6-1) is given as follows, in the same way as in Chapter 5.

*'Matter $\omega_1$ causes Attribute $\omega_5$ of Matter $\omega_2$ to keep ($\omega_3 = \omega_4$) or change ($\omega_3 \neq \omega_4$) its Values temporally ($\omega_6 = G_t$) or spatially ($\omega_6 = G_s$) over a certain absolute time-interval, where Values $\omega_3$ and $\omega_4$ are relative to Standard $\omega_7$.'*

It is notable that Matter terms placed at $\omega_3$, $\omega_4$ or $\omega_7$ represent their values in each place at the time or over the time-interval. In this sense, (6-1) is an augmented version of (5-1) in order to cope with surrogate expressions at value or standard such as S6-1 and S6-2 instead of S6-3 and S6-4, respectively.

(S6-1) Tom is at *home*.

(S6-2) Tom is taller than *Jim*.

(S6-3) Tom is now *at 40° north 135° east*.

(S6-4) Tom is taller than *6 feet*.

Hereafter, when it is not significant to discern variables at $\omega_1$, $\omega_3$, $\omega_4$ or $\omega$, anonymous variables are often inserted, usually symbolized as '_' for the sake of simplicity. In particular, this special symbol is often placed instead of an event causer ($\omega_1$) or Standard variable ($\omega_7$) because of no significance for people to consider or reason. Furthermore, the notation (∃...) is often employed in order to imply that the otherwise unbounded variables are bound by ∃.

## 6.3 Tempo-Logical Connectives

Robots must be provided with every kind of knowledge, especially knowledge about space and time, such as causal chains in the physical world. For example, such an assertion as S6-5 does not refer to a special event such as S6-6 but a certain law belonging to our common sense knowledge.

(S6-5) The air gets warmer after the sun rises.

(S6-6) The air got warmer after the sun rose.

In order to facilitate formal representation and computation of such knowledge, the deductive system employs tempo-logical connectives, defined as $\mathbf{D}_{\mathbf{TL}}$. X⇔Y reads 'X is defined by Y'. The definition of a tempo-logical connective $\Theta_i$ is simply a generalization of that for tempo-logical conjunctions ($\wedge_i$) given by (5-4) only with such difference that $\Theta$ refers to every binary connective in conventional logic, namely, AND ($\wedge$), OR ($\vee$), IMPLICATION ($\supset$), or EQUIVALENCE ($\equiv$). The definition of each $\tau_i$ is all the same as that given in Chapter 5. As already mentioned in the previous chapter, the tempo-logical connectives used most frequently are SAND ($\wedge_0$) and CAND ($\wedge_1$), standing for *Simultaneous AND* and *Consecutive AND* and conventionally symbolized as $\Pi$ and $\bullet$, respectively. As for operation order, tempo-logical connectives are supposed to precede conventional logical connectives.

$\mathbf{D}_{\mathbf{TL}}$. $\chi_1\ \Theta_i\ \chi_2 \Leftrightarrow (\chi_1\ \Theta\ \chi_2) \wedge \tau_i(\chi_1, \chi_2)$,

where $\Theta \in \{\wedge, \vee, \supset, \equiv\}$ and $\tau_{-i}(\chi_2, \chi_1) \equiv \tau_i(\chi_1, \chi_2)$ ($\forall i \in \{0,\pm1,\pm2,\pm3,\pm4,\pm5,\pm6\}$).

In order for further expansion of $\boldsymbol{L}_{md}$, the definition of Empty Event (5-7) is redefined as $D_{EE}$ below, representing Temporal Empty Event for $g$=*Gt* and Spatial Empty Event for $g$=*Gs*. These two kinds of empty events imply independent movements of the focus of attention of the observer in time and in space, respectively. The usage of spatial empty event is to be given later.

$\mathbf{D}_{\mathbf{EE}}$. $\varepsilon([t_i,t_j]) \Leftrightarrow (\exists g)L(Time,Time,t_i,t_j,A_{34},g,S_{Ta})$,

where $[t_i,t_j] \in \Delta$ (=$\{[p,q] \mid p \leq q$, p and q are real numbers$\}$).

## 6.4 Formulation of Event Concepts

Event concepts expressed by verbs, for example, are defined as generalized locus patterns. For example, the concepts of the English verbs *carry* and *return* are to be defined as (6-2) (or (6-2') with abbreviation at Attribute Carriers) and (6-3), respectively. The special symbol $\lambda$ is employed to rigidly define relations or functions, implying that its following symbols are free variables. These formulas can be depicted as Fig. 6-1a and b, respectively.

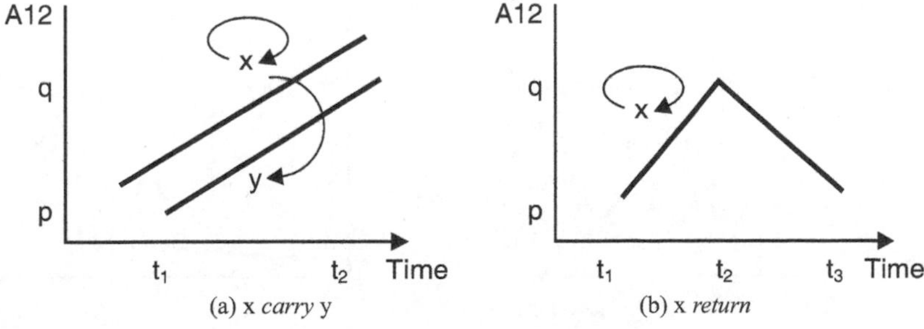


(a) x *carry* y (b) x *return*

**Fig. 6-1.** Loci of *carry* and *return*.

$$(\lambda x,y)carry(x,y) \Leftrightarrow (\lambda x,y)(\exists p,q,k)L(x,x,p,q,A_{12},G_t,k)\Pi L(x,y,p,q,A_{12},G_t,k) \wedge x\neq y \wedge p\neq q \quad (6\text{-}2)$$

$$\Leftrightarrow (\lambda x,y)(\exists p,q,k)L(x,\{x,y\},p,q,A_{12},G_t,k) \wedge x\neq y \wedge p\neq q \quad (6\text{-}2')$$

$$(\lambda x)return(x) \Leftrightarrow (\lambda x)(\exists p,q,k)L(x,x,p,q,A_{12},G_t,k) \bullet L(x,x,q,p,A_{12},G_t,k) \wedge x\neq y \wedge p\neq q \quad (6\text{-}3)$$

The expression (6-4) is the definition of the English verb concept *fetch*, depicted as Fig. 6-2a. This implies such a temporal event that *x* goes for *y* and then comes back with it.

$$(\lambda x,y)fetch(x,y) \Leftrightarrow (\lambda x,y)\ (\exists p_1,p_2,k)\ L(x,x,p_1,p_2,A_{12},G_t,k) \bullet ((\underline{L(x,x,p_2,p_1,A_{12},G_t,k)\Pi L(x,y,p_2,p_1,A_{12},G_t,k)}))\ \wedge x\neq y \wedge p_1\neq p_2 \quad (6\text{-}4)$$

In the same way, the English verb concept *hand* or *receive*, depicted in Fig. 6-2b, is defined equivalently as (6-5) or its abbreviation (6-5') where Event Causers are merged into a set.

$$(\lambda x,y,z)hand(x,y,z).\equiv.(\lambda x,y,z)receive(z,y,x) \quad (6\text{-}5)$$

$$\Leftrightarrow (\lambda x,y,z)(\exists k)L(x,y,x,z,A_{12},G_t,k)\Pi L(z,y,x,z,A_{12},G_t,k) \wedge x\neq y \wedge y\neq z \wedge z\neq x$$

$$\Leftrightarrow (\lambda x,y,z)(\exists k)L(\{x,z\},y,x,z,A_{12},G_t,k) \wedge x\neq y \wedge y\neq z \wedge z\neq x \quad (6\text{-}5')$$

Such locus formulas as correspond with natural event concepts are called *Event Patterns*, and about 40 kinds of event patterns have been found concerning the attribute *Physical location* $(A_{12})$, for example, *start*, *stop*, *meet*, *separate*, *carry*, *return*, etc., shown in Fig. 6-3 (e.g., Yokota, 2005).

In the author's mind, such verbs as *meet* and *separate* can refer to both spatial and temporal change events as well as *run*, which implies that the pseudo constant $G_x$ takes either $G_s$ or $G_t$ correspondingly in (6-6) and (6-7). The variable *z* refers to a certain event causer.

$$(\lambda x,y)meet(x,y) \Leftrightarrow (\lambda x,y)(\exists z,p,q,k)L(z,x,p,r,A_{12},G_x,k)\Pi L(z,y,q,r,A_{12},G_x,k) \wedge x\neq y \wedge p\neq q \wedge q\neq r \wedge r\neq p \quad (6\text{-}6)$$

$$(\lambda x,y)separate(x,y) \Leftrightarrow (\lambda x,y)(\exists z,p,q,k)L(z,x,r,p,A_{12},G_x,k)\Pi L(z,y,r,q,A_{12},G_x,k)\ \wedge x\neq y \wedge p\neq q \wedge q\neq r \wedge r\neq p \quad (6\text{-}7)$$

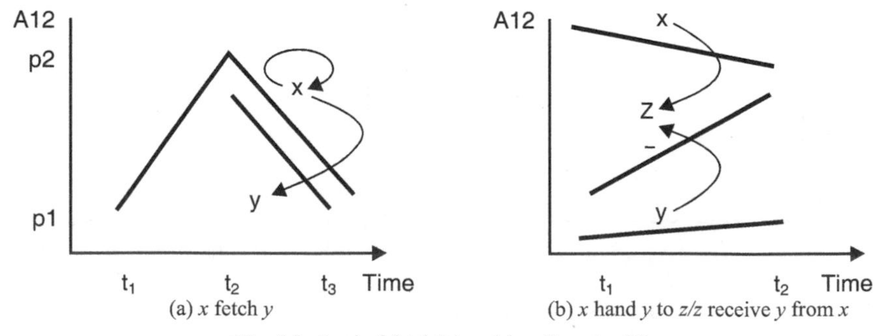


(a) *x* fetch *y* (b) *x* hand *y* to *z*/*z* receive *y* from *x*

**Fig. 6-2.** Loci of *fetch* (a) and *hand*/*receive* (b).

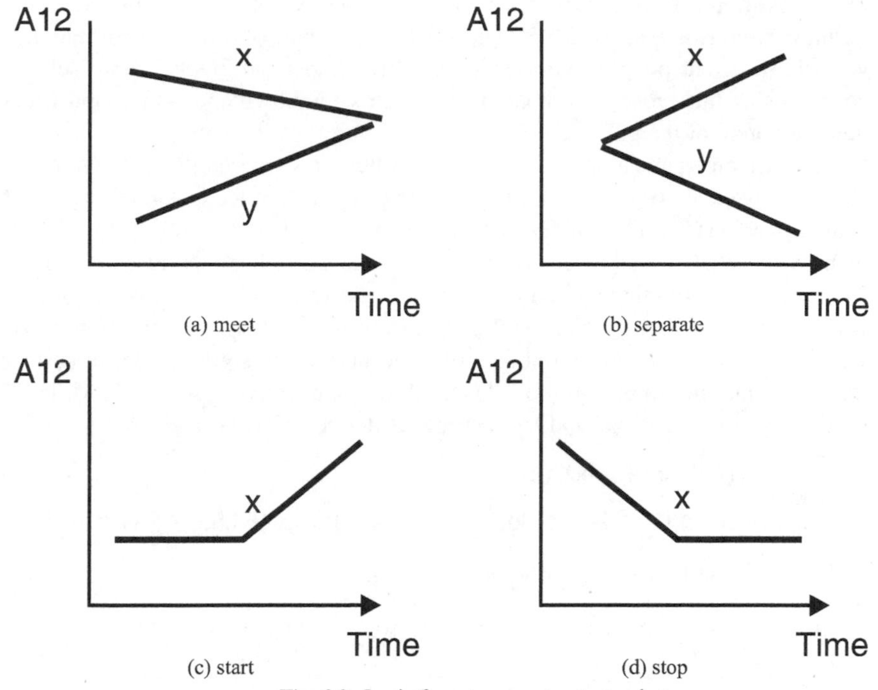


**Fig. 6-3.** Loci of *meet, separate, start* and *stop.*

## 6.5 Formulation of Matter Concepts

A matter, usually referred to by a noun in natural language, is to be conceptualized as a conjunction of the mental images of itself and its relations with others that, in turn, are to be reduced to certain loci in attribute spaces.

A matter concept is a generalized integration of events about all the attributes, implying how the very matter does concern them or not. In the formal system, a matter concept '$\psi$' is introduced in such a context as (6-8), where '$\psi^{+}$' and '$\psi^{++}$' are to represent the conceptual images of itself and its relations with others, respectively, and to be reduced to atomic locus formulas of all the attributes. The two kinds of conceptual images are respectively related to *Merkbild* (image for recognition) and *Wirkbild* (image for workout) in the field of biology (Uexküll and Kriszat, 1934). For example, an animal is to recognize a prey by its Merkbild (e.g., its shape) and capture it according to a certain attack pattern conceptualized as its Wirkbild (e.g., biting it).

$$(\lambda z)\psi(z)\Leftrightarrow(\lambda z)\psi^{+}(z)\wedge\psi^{++}(z) \qquad (6\text{-}8)$$

Whereas $\psi(z)$ must be a total description of all the attributes, for simplicity only its important part is to be given here with the symbol % representing its abbreviated part. The part $\psi^{+}(z)$ is given as a combination of atomic locus formulas for the Attribute Carrier *z* without any other specific matter designated at the Event Causer as is indicated by '_' unlike the other part $\psi^{++}(z)$. This corresponds with the fact that people live their casual life almost intuitively, being seldom conscious of the causes of phenomena such as rainbows and winds and often being ignorant of them.

For example, the matter called *ice* can be not *scientifically* but *intuitively* conceptualized as (6-9). This formula as for $\psi^{+}$ reads that ice is something *x* that is always cold (*Cld*, $A_{28}$: *Temperature*) and solid (*Sld*, $A_{41}$: *Quality*), that is always of no vitality ($A_{39}$), and as for $\psi^{++}$ that (at some time) happens to change into/from water (or from that something else *z* (scientifically, $H_2O$) changes into/from water), and ...'. The special symbol '_', defined by (6-10), is an anonymous variable bound by an existential quantifier because of no significance for further reasoning and in turn the symbol '/' is used for the negative case of '_'. while '*' and '◊' represent 'always' and 'some time' as defined by (6-11).

$$(\lambda x)ice(x)\Leftrightarrow(\lambda x)ice^{+}(x)\wedge ice^{++}(x),$$

$$(\lambda x)ice^{+}(x)\Leftrightarrow(\lambda x)((\exists\ldots)L(\_,x,Cld,Cld,A_{28},G_t,k_1)\Pi L(\_,x,Sld,Sld,A_{41},G_t,k_2))^{*}\wedge$$

$$(\sim(\exists\ldots)L(\_,x,p,p,A_{39},G_t,k_3))^{*}\wedge\%,$$

$$(\lambda x)ice^{++}(x)\Leftrightarrow((\exists\ldots)L(x_1,z,x,x_2,A_{41},G_t,k_1))^{\Diamond}\wedge((\exists\ldots)L(x_3,z,x_2,x,A_{41},G_t,k_1))^{\Diamond}\wedge water(x_2)\wedge\% \quad (6\text{-}9)$$

$$L(\ldots,\omega_i,\_,\omega_j,\ldots)\Leftrightarrow(\exists\omega)L(\ldots,\omega_i,\omega,\omega_j,\ldots),\quad L(\ldots,\omega_i,/,\omega_j,\ldots)\Leftrightarrow\sim L(\ldots,\omega_i,\_,\omega_j,\ldots). \quad (6\text{-}10)$$

$$\chi^{*}\Leftrightarrow(\forall\mathbf{d})\,\chi\,\Pi\,\varepsilon(\mathbf{d}),$$

$$\chi^{\Diamond}\Leftrightarrow(\exists\mathbf{d})\,\chi\,\Pi\,\varepsilon(\mathbf{d}), \quad (6\text{-}11)$$

where $\mathbf{d}\in\Delta$ (=$\{[p,q] \mid p\leq q$, p and q are real numbers$\}$).

In turn, the matter *water*, as defined by (6-12), is something $x_2$ that is liquid (*Lqd*, $A_{41}$: *Quality*) that (at some time) happens to change into/from ice, and that human happens to take from (own) mouth to (own) stomach, ... where the underlined part denotes that the mouth and the stomach are at the human, where the attribute $A_{12}$ refers to *Physical Location*.

$$(\lambda x_2)water(x_2)\Leftrightarrow((\exists\ldots)L(y,x_2,Lqd,Lqd,A_{41},G_t,k))^{*}\wedge((\exists\ldots)L(y,x_2,x_2,x,A_{41},G_t,k_1))^{\Diamond}\wedge$$

$$((\exists\ldots)L(y,x_2,x,x_2,A_{41},G_t,k_1))^{\Diamond}\wedge ice(x)\wedge((\exists\ldots)L(z,x_2,z_1,z_2,A_{12},G_t,k_2)$$

$$\Pi L(\underline{z,\{z_1,z_2\},z,z,A_{12},G_t,k_2}))^{\Diamond}\wedge human(z)\wedge mouth(z_1)\wedge stomach(z_2)\wedge\%. \quad (6\text{-}12)$$

For another example, the matter *snow* can be conceptualized as (6-13), reading 'Snow is something *x* that is always powdered ice, that happens to be attracted from sky by earth, and ...'.

$$(\lambda x)snow(\lambda x)\Leftrightarrow(x)((\exists\ldots)L(\_,x,x_1,x_1,A_{41},G_t,k_1)\Pi L(\_,x,x_2,x_2,A_{41},G_t,k_2))^{*}$$

$$\wedge\,(L(Earth,x,Sky,Earth,A_{12},G_t,k_3))^{\Diamond}\wedge powder(x_1)\wedge ice(x_2)\wedge\%. \quad (6\text{-}13)$$

As easily understood, matter concepts include miscellaneous spatiotemporal relations (i.e., events) among matters usually referred to by verbs, prepositions, etc. Therefore, a matter concept is usually much more complicated than an event concept in definition. By the way, a noun originated or derived from a verb is to be defined as a matter concept in the same form as (6-8), including the very verb concept as for $\psi^{++}$. For example, the concept of 'conveyance' is to be introduced as (6-14), where the definition of *convey(x,y)* is to be given by (6-2), the same as *carry(x,y)* (if no special distinction is required) and $A_{32}$ is the attribute Color.

$$(\lambda z)\text{conveyance}(z) \Leftrightarrow (\lambda z)\text{conveyance}^{+}(z) \wedge \text{conveyance}^{++}(z),$$
$$(\lambda z)\text{conveyance}^{+}(z) \Leftrightarrow (\lambda z)(\sim(\exists\ldots)L(\_,z,p,q,A_{32},G_t,k))^{*} \wedge \%,$$
$$(\lambda z)\text{conveyance}^{++}(z) \Leftrightarrow (\lambda z)(\exists\ldots)L(\_,\{x,y\},z,z,A_{12},G_t,k)\Pi \text{convey}(x,y). \qquad (6\text{-}14)$$

According to this formalization of a matter concept, for example, the two expressions S6-7 and S6-8 are to be translated into the same expression (6-15) in $\boldsymbol{L}_{md}$. The attribute $A_{16}$ is for Velocity and the value for *quickness* is indicated by the formula, $r > k_1$.

(S6-7) Amazon's conveyance of goods is quick.

(S6-8) Amazon conveys goods quickly.

$$(\exists\ldots)L(\_,\{\text{Amazon},y\},z,z,A_{12},G_t,k)\Pi L(\_,z,r,r,A_{16},G_t,k_1)\Pi \text{convey}(\text{Amazon},y) \wedge r > k_1 \wedge \text{goods}(y) \qquad (6\text{-}15)$$

The variable *z* corresponds with the variable *e* in Davidson's logical form (Davidson, 1967), introduced to formalize an event in first order predicate logic (e.g., Hobbs, 1985). As easily understood, in mental image directed semantic theory, the atomic locus formulas, including the variable *z*, can be logically eliminated in the case of no use or significance. Cognitively, this variable refers to a certain psychological existence called Imaginary Space Region which can be created arbitrarily by a human, as shown in Fig. 6-4 (c.f., the postulate of arbitrary

$$(\lambda z)\ \text{conveyance}(z) \Leftrightarrow (\lambda z)\ (\exists\ldots)\ L\ (\_,\{x,y\},z,z,A12,Gt,\_)\ \Pi\ L(x,x,p,q,A12,Gt,\_)\Pi\ L(x,y,p,q,A12,Gt,\_) \wedge x\neq y \wedge p\neq q$$

**Fig. 6-4.** Concept formation of conveyance by human.

production of imaginary space region at 7.1.7), and logically the atomic locus formulas concerning it are assumed to hold any time.

Viewed from natural language understanding, a matter concept is a combination of its potential events involved at every attribute to be employed for semantic processing, for example, semantic anomaly detection in such adjunct expressions as '*hot snow*' and '*…convey colorfully…*'.

## 6.6 Formulation of Laws of the World

Various kinds of laws in the daily life are to be formalized by employing tempo-logical implications (i.e., $\supset_i$) because they are implicational relations between two events in different time-intervals, such as that expressed by S6-9 and the social rule expressed by S6-10, roughly formulated as (6-16) and (6-17) in $\boldsymbol{L}_{md}$, respectively. For another example, S6-11 should be translated into (6-18) but S6-12 into (6-19).

(S6-9) If it rains, the ground will get wetter.

(S6-10) If the traffic signal is red at a crossing, we must not proceed onto the crossing.

$$\text{Rain} \supset_5 \text{Get\_wetter(Ground)} \qquad (6\text{-}16)$$

$$\text{Red(T-signal)} \supset_{-3} \sim\text{Proceed(We)} \qquad (6\text{-}17)$$

(S6-11) The air got warmer after the sun rose.

(S6-12) The air gets warmer after the sun rises.

$$\text{Rise(Sun)} \wedge_5 \text{Get\_warmer(Air)} \qquad (6\text{-}18)$$

$$\text{Rise(Sun)} \supset_5 \text{Get\_warmer(Air)} \qquad (6\text{-}19)$$

This can be understood by the fact that S6-11 can be rephrased as S6-13 while S6-12 should not be restated as S6-14 but S6-15 to be interpreted as (6-20) in $\boldsymbol{L}_{md}$.

(S6-13) The sun rose before the air got warmer.

(S6-14) The sun rises before the air gets warmer.

(S6-15) If the air does not get warmer, the sun does not rise in advance.

$$\sim\text{Get\_warmer(Air)} \supset_{-5} \sim\text{Rise(Sun)} \qquad (6\text{-}20)$$

In general, the following equivalence relation (6-21) holds. This is the theorem of tempo-logical contrapositive whose proof is to be given later.

$$\chi_1 \supset_i \chi_2 .\equiv. \sim\chi_2 \supset_{-i} \sim\chi_1 \qquad (6\text{-}21)$$

It is noticeable that tempo-logical implications are very useful for formulating tempo-logical relationships between miscellaneous event concepts without explicit indication of time intervals. For example, as understood intuitively by Fig. 6-1a and Fig. 6-2a, and rigidly by the formulas (6-2) and (6-4), an event *fetch*(*x,y*) is necessarily *finished by* an event *carry*(*x,y*). This fact can be formulated as (6-22), where $\supset_{-4}$ is $\supset$ furnished with the temporal relation *finished-by* (i.e., $\tau_{-4}$). This kind of formula is not an axiom but a theorem deducible from the definitions of event concepts in our formal system.

$$(\forall x,y)(\text{fetch}(x,y) \supset_{-4} \text{carry}(x,y)) \quad (6\text{-}22)$$

In the same way, we can also know that *fetch*(*x,y*) simultaneously involves *return*(*x*) by (6-3) as explicated by (6-23).

$$(\forall x,y)(\text{fetch}(x,y) \supset_{0} \text{return}(x)) \quad (6\text{-}23)$$

# 7
# Fundamental Postulates and Inference Rules for Deductive System

It is most crucial to provide the natural language understanding system with certain metaphysics or ontology of mental image in order for systematic mental image computation. All knowledge pieces resulting from an individual's everyday experience are inevitably subjective (to him/her), that is, not necessarily intelligible to others. In this sense, the formal system is subjective (to the author) as far as it employs (domain-specific) constants other than logical ones, such as logical connectives (generally assumed to be objective).

Moreover, such knowledge pieces are not always valid even for their holder. For example, a word concept should be associated with such a conceptual image that it is abstract enough to represent the perceptual image of every referent of the word. However, it is practically impossible for an individual to obtain a conceptual image in this sense because instances or referents of a concept are usually too numerous for him/her to encounter and observe. Therefore, our conceptual image for a word is always imperfect or tentative to be occasionally updated by an exceptional instance just like a 'black swan'. Therefore, such definitions of word concepts introduced above should not be employed as axioms but as hypothetical knowledge pieces.

This chapter focuses on (Yokota's) subjective or empirical laws, called *postulates*, and tempo-logical inference rules to be applied to all $\boldsymbol{L_{md}}$ expressions. The postulates are of all considerable attributes (physical or not), although all the examples for their applications here concern the physical world in order for better comprehension. They are to be treated as equivalents to axioms as long as they do not encounter any exception.

## 7.1 Properties of Loci

### *7.1.1 Identity of assigned values*

The postulates denoted as $\mathbf{P}_{V1}$ and $\mathbf{P}_{V2}$ below state that ***a matter never has different values of an attribute with a standard at a time.*** These are called *Postulate of Identity of Assigned Values-type1 and Postulate of Identity of Assigned Values-type2, respectively.* Hereafter, (∀...) implies that the otherwise unbounded variables are bound by ∀.

$\mathbf{P}_{V1}$. $(\forall\ldots)L(x,y,p_1,q_1,a,G_t,k)\Pi L(z,y,p_2,q_2,a,G_t,k).\supset.p_1=p_2\wedge q_1=q_2.$

$\mathbf{P}_{V2}$. $(\forall\ldots)L(x,y,p_1,q_1,a,G_t,k)\bullet L(z,y,p_2,q_2,a,G_t,k).\supset.q_1=p_2.$

$\mathbf{P}_{V1}$ is employed exclusively to detect semantic anomaly in such a sentence as S7-1 while $\mathbf{P}_{V2}$ is useful to detect event gaps in such a context as S7-2, leading to the inferential consequent commented in the parentheses below.

(S7-1) The red box is black.

(S7-2) Taro was in Tokyo yesterday and he is in London today. (Therefore, Taro travelled from Tokyo to London.)

The syntax of $L_{md}$ allows Matter terms to appear at Value and Standard in order to represent their values in each place at the time and over the time-interval, respectively. This rule can be formulated as Postulate of Matter as Value and Postulate of Matter as Standard, denoted as $P_{MV}$ and $P_{MS}$ below, respectively. For example, the value of Physical Location ($A_{12}$) of reference by a matter term $x$ is called psychologically occupied region of $x$, which could be referred to by the English preposition 'at', implying 'in', 'on', or 'near' (c.f. 8.3). This assumption is to be valid for the other attributes as its analogies.

$\mathbf{P}_{MV}$. $(\forall\ldots)L(x_0,y,z_1,z_2,a,g,k)\Pi L(x_1,z_1,p_1,q_1,a,g,k)$
$\Pi L(x_2,z_2,p_2,q_2,a,g,k).\supset_0.\ L(x_0,y,p_1,q_2,a,g,k).$
If $z_1=z_2$, this is reduced to the formula below.
$(\forall\ldots)L(w,y,x,x,a,g,k)\Pi L(z,x,p,q,a,g,k)\ .\supset_0.\ L(w,y,p,q,a,g,k).$

$\mathbf{P}_{MS}$. $(\forall\ldots)L(x_0,y,p_1,p_2,a,g,z)\Pi L(x_1,z,q,q,a,g,k).\supset_0.L(x_0,y,p_1,p_2,a,g,q).$

$\mathbf{P}_{MV}$ is to be utilized for such inference as S7-3, while $\mathbf{P}_{MS}$ is for such inference as S7-4.

(S7-3) The letter was sent from Tom to Mary. Tom was in Tokyo. Mary was in Osaka. Then, the letter moved from Tokyo to Osaka.

(S7-4) Jim is taller than Tom. Tom is 2 m tall. Therefore, Jim is taller than 2 m.

### 7.1.2 Arbitrariness of Locus Articulation

Articulating a locus is quite subjective. For example, whether the point ($t_2$, q) in Fig. 7-1a is significant or not so, as in Fig. 7-1b, more generally, locus articulation depends on the precisions or the granularities of these standards, which can be formulated as the postulates denoted as $\mathbf{P}_{A1}$ and $\mathbf{P}_{A2}$, called ***Postulate of Arbitrariness in Locus Articulation-type1 and Postulate of Arbitrariness in Locus Articulation-type2, respectively***. These postulates affect the process of conceptualization on a word based on its referents in the world and, moreover, they are very useful for spatiotemporal inference.

$\mathbf{P}_{A1}$. $(\forall\ldots)(\exists k')L(y,x,p,q,a,g,k)\bullet L(y,x,q,r,a,g,k).\supset_0.L(y,x,p,r,a,g,k')$.

$\mathbf{P}_{A2}$. $(\forall\ldots)(\exists r,k')L(y,x,p,r,a,g,k).\supset_0.L(y,x,p,q,a,g,k')\bullet L(y,x,q,r,a,g,k')$.

For example, $\mathbf{P}_{A1}$ and $\mathbf{P}_{A2}$ are applicable to such reasoning as S7-5 and as S7-6, respectively.

(S7-5) Tom flied from Tokyo to Nagoya and successively to Osaka. Therefore, he moved from Tokyo to Osaka.

(S7-6) Tom moved from Tokyo to Osaka. Therefore, he passed somewhere (between the two places on the way).

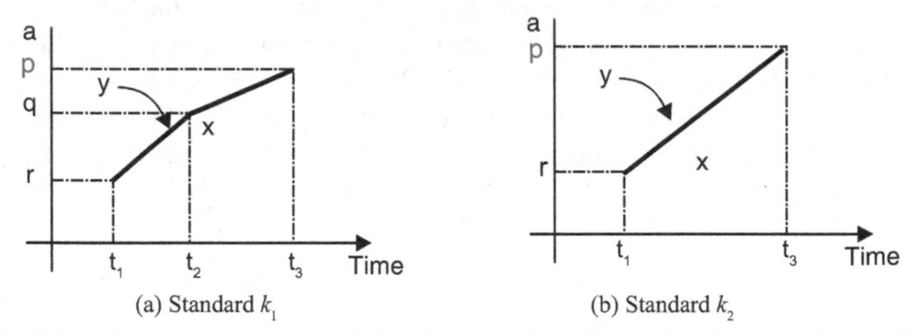


(a) Standard $k_1$ (b) Standard $k_2$

**Fig. 7-1.** Arbitrariness in locus articulation due to standards (Standard $K_1$ (a) is finer than $K_2$ (b)).

### 7.1.3 Negation-freeness of absolute time passing

Loci for perceptual images corresponding to real live scenes are called perceptual loci. A perceptual locus can be formulated only with atomic locus formulas and temporal conjunctions, called perceptual locus formula. This is not necessarily the case for locus formulas of conceptual images, called conceptual loci, corresponding to certain generalized knowledge pieces. For example, people usually interpret the construction '*B* happens *before/after A* happens' as a general causality, namely, as 'If *A* happens, *B* happens *in advance/later*.' More concretely, consider S7-7.

(S7-7) Tom washes his face immediately after he gets up (every morning).

This kind of knowledge should be formulated with temporal implications in such a way as $A \supset_i B$. However, the definition of tempo-logical connectives (i.e., $D_{TL}$ introduced in Chapter 6) is exclusively for a real live image to be formulated by positive atomic loci and temporal *conjunctions*. That is, $D_{TL}$ holds on such a presumption that the absolute time intervals of a locus formula and its negation are identical because, according to standard logic, $A \supset B$ is equivalent to $\sim A \vee B$. However, there is no interpreting a negated locus formula as a locus with ***a unique time-interval*** to determine a unique temporal relation $\tau_i$.

Considering such a theorem as $\sim(A \vee B) .\equiv. \sim A \wedge \sim B$ in standard logic, it is quite natural to assume identity of a locus formula with its negative in absolute time-interval, that is, negation-freeness of absolute time passing. Therefore, in order to also make definition of empty event valid for conceptual image, we introduce a meta-function $\delta$, defined by Definition of Time-interval and its related postulates, Postulate of Negation of Time-interval and Postulate of Compound Time-interval denoted as $D_{TI}$, $P_{NT}$ and $P_{CT}$, respectively, as follows, where $\delta$ is to extract the ***suppressed*** absolute interval of a locus formula $\chi$.

**$D_{TI}$.** $\delta(\chi)=[t_a,t_b](\in\Delta)$,
where $\chi\Pi\varepsilon([t_a,t_b])$.

**$P_{NT}$.** $\delta(\sim\alpha)=\delta(\alpha)$,
where $\alpha$ is an atomic locus formula.

**$P_{CT}$.** $\delta(\chi)=[t_{min}, t_{max}]$,

where $t_{min}$ and $t_{max}$ are respectively the *minimum* and the *maximum* time-point included in the absolute time-intervals of the atomic locus formulas, either positive or negative, within $\chi$.

These postulates lead to the Theorem of Absoluteness of Time Passing (or Negation-freeness of Absolute Time Passing) denoted as $T_{TP}$ below (Yokota, 2008). This theorem can read that absolute time passes during an *objective* event whether it may be perceived *subjectively* as $\chi$ or as $\sim\chi$.

**$T_{TP}$.** $\delta(\sim\chi)=\delta(\chi)$.

(Proof) According to $\mathbf{P_{NT}}$ and $\mathbf{P_{CT}}$, the time-interval of each atomic locus formula involved in $\sim\chi$ is negation-free and therefore so are $t_{min}$ and $t_{max}$ in $\delta(\sim\chi)$. **[Q.E.D.]**

The counterpart of the contrapositive in standard logic (i.e., $A \supset B .\equiv. \sim B \supset \sim A$) is given as Theorem of Tempo-logical Contrapositive ($\mathbf{T_{TC}}$) (Yokota, 2008) whose rough proof is as follows immediately below, where the left hand of ':' refers to the theses, such as theorems (e.g., **PL** denotes a subset of those in (pure) predicate logic) employed at the bi-directional deduction indicated by the meta-symbol '↔' (while '→' for one-directional deduction).

**$T_{TC}$.** $\chi_1 \supset_i \chi_2 .\equiv. \sim\chi_2 \supset_{-i} \sim\chi_1$.

(Proof)

$\mathbf{D}_{\mathbf{TL}}$: $\chi_1 \supset_i \chi_2 \leftrightarrow (\chi_1 \supset \chi_2) \wedge \tau_i(\chi_1, \chi_2)$

**PL:** $\leftrightarrow (\sim\chi_2 \supset \sim\chi_1) \wedge \tau_i(\chi_1, \chi_2)$

$\mathbf{T}_{\mathbf{TP}}$: $\leftrightarrow (\sim\chi_2 \supset \sim\chi_1) \wedge \tau_i(\sim\chi_1, \sim\chi_2)$

$\mathbf{D}_{\mathbf{TL}}$: $\leftrightarrow (\sim\chi_2 \supset \sim\chi_1) \wedge \tau_{-i}(\sim\chi_2, \sim\chi_1)$

$\mathbf{D}_{\mathbf{TL}}$: $\leftrightarrow \sim\chi_2 \supset_{-i} \sim\chi_1$

**[Q.E.D.]**

Therefore, S7-8 and S7-9 are proved to be paraphrases of each other by employing $\mathbf{T}_{\mathbf{TC}}$ while S7-10 and S7-11 are proved so by the definition of tempological conjunctions (i.e., $\wedge_i$).

(S7-8) It gets cloudy *before* it rains.
=If it rains, it gets cloudy *in advance*. (≡Rain $\supset_{-5}$ Get_Cloudy)

(S7-9) It does not rain *after* it does not get cloudy.
=Unless it gets cloudy, it does not rain *later*. (≡~Get_Cloudy $\supset_5$ ~Rain)

(S7-10) It got cloudy *before* it rained. (≡Rain$\wedge_{-5}$Get_Cloudy)

(S7-11) It rained *after* it got cloudy. (≡Get_Cloudy$\wedge_5$Rain)

Any event always coexists with an empty event as is formulated by Theorem of Coexistence of Empty Event denoted as $\mathbf{T}_{\mathbf{CE}}$ below whose proof is trivial.

$\mathbf{T}_{\mathbf{CE}}.$ $\chi .\equiv_0. \chi\ \Pi\varepsilon(\delta(\chi))$

### *7.1.4 Shortcut in Causal Chain*

$\mathbf{P}_{\mathbf{SC}}.$ $(\forall\ldots)L(z,x,p,q,a,g,k)\Pi L(w,y,x,x,a,g,k)\ .\supset_0.\ L(z,x,p,q,a,g,k)$
$\Pi L(z,y,p,q,a,g,k).$

'This postulate is called *Postulate of Shortcut of Causal Chain*. When $a=A_{12}$ (*Physical location*) and $p \neq q$, for example, $\mathbf{P}_{\mathbf{SC}}$ reads that if *z* causes *x* to move from *p* to *q* while *w* causes *y* to be with *x* then *w* causes *y* to move from *p* to *q*. This postulate can be depicted as Fig. 7-2.

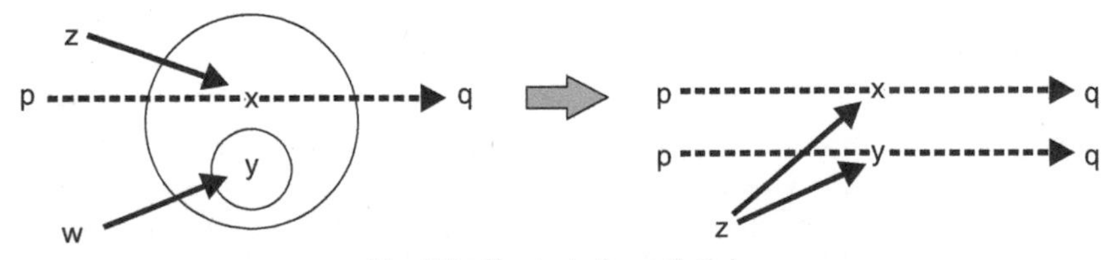


**Fig. 7-2.** Shortcut of causal chain.

### 7.1.5 Conservation of Values

$\mathbf{P}_{CV}$. $(\forall\ldots)L(z,x,p,p,a,g,k)\bullet\chi .\supset_0. L(z,x,p,p,a,g,k)\bullet(\underline{L(z,x,p,p,a,g,k)\Pi\chi})$
if and only if the underlined part does not violate the poslulate $\mathbf{P}_{A1}$.

This is called *Postulate of Conservation of Values*. For example, when $a=A_{12}$ (*Physical location*), $\mathbf{P}_{CV}$ reads that if *z* causes *x* to be at *p* then *z* causes *x* to be at *p* successively. Note that $\mathbf{P}_{CV}$ is conditional because it is predictive. That is, it is valid only when $\chi$ does not contradict with *L(z,x,p,p,a,g,k)*.

### 7.1.6 Reversibility of Spatial Change Events

As already mentioned in 4.2, all loci in attribute spaces are assumed to correspond one to one with movements or, more generally, temporal change events of the focus of attention of the observer. Therefore, the $L_{md}$ expression of an event is compared to a movie recorded through a floating camera because it is necessarily grounded in movement of the focus of attention of the observer over the event. This is why S7-12 and S7-13 can refer to the same scene in spite of their appearances, where what 'sinks' or 'rises' is the focus of attention of the observer as illustrated in Fig. 7-3a. These are interpreted into the $L_{md}$ expressions (7-1) and (7-2), respectively, where '$A_{13}$', '↑' and '↓' refer to the attribute 'Direction' and its values 'upward' and 'downward', actually represented as unit vectors, respectively.

(S7-12) The path sinks to the brook.

(S7-13) The path rises from the brook.

$$(\exists\ldots)L(x,y,z_1,z_2,A_{12},G_s,k)\Pi L(x,y,\downarrow,\downarrow,A_{13},G_s,k)\wedge path(y)\wedge brook(z_2)\wedge z_1\neq z_2. \quad (7\text{-}1)$$

$$(\exists\ldots)L(x,y,z_2,z_1,A_{12},G_s,k)\Pi L(x,y,\uparrow,\uparrow,A_{13},G_s,k)\wedge path(y)\wedge brook(z_2)\wedge z_1\neq z_2. \quad (7\text{-}2)$$

Such a fact is generalized as the postulate $\mathbf{P}_{RS}$ called *Postulate of Reversibility of Spatial Change Event*, where $\chi_s$ and $\chi_s^R$ are a perceptual locus and its *reversal* for a certain spatial change event, respectively, and they are substitutable with each other because of the property of $\equiv_0$. This postulate can be one of the principal inference rules belonging to peoples' common-sense knowledge about geography. Not to mention, this postulate does not hold for temporal change events.

$\mathbf{P}_{RS}$. $\chi_s^R .\equiv_0. \chi_s$

This postulate is also valid for such a pair of S7-14 and S7-15 as interpreted approximately into (7-3) and (7-4), respectively, referring to the same scene as shown in Fig. 7-3b. These pairs of conceptual descriptions are called equivalent in $\mathbf{P}_{RS}$, and the paired sentences are treated as paraphrases of each other.

(S7-14) Route A and Route B meet at the city.

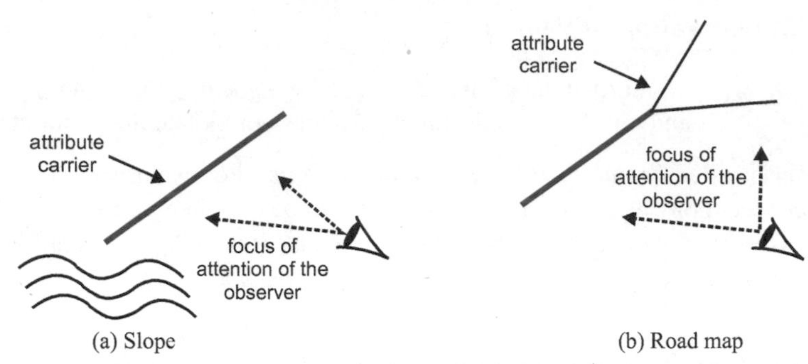


**Fig. 7-3.** Spatial change events and the focus of attention of the observer movements.

(S7-15) Route A and Route B separate at the city.

$$(\exists\ldots)L(x,Route\_A,z_1,y,A_{12},G_s,k)\Pi L(x,Route\_B,z_2,y,A_{12},G_s,k)\wedge city(y)\wedge z_1\neq z_2 \quad (7\text{-}3)$$

$$(\exists\ldots)L(x,Route\_A,y,z_1,A_{12},G_s,k)\Pi L(x,Route\_B,y,z_2,A_{12},G_s,k)\wedge city(y)\wedge z_1\neq z_2 \quad (7\text{-}4)$$

The postulate $\mathbf{P}_{\mathbf{RS}}$ is accompanied with the mental operation *reversal*, denoted by $R$ and recursively defined as $\mathbf{D}_{\mathbf{RO}}$ called Definition of Reversal Operation of Spatial Change Event, where $\chi_i$ stands for a locus formula.

$\mathbf{D}_{\mathbf{RO}}$.
$$(\chi_1\bullet\chi_2)^R\Leftrightarrow\chi_2^R\bullet\chi_1^R$$
$$(\chi_1\Pi\chi_2)^R\Leftrightarrow\chi_1^R\Pi\chi_2^R$$
$$L^R(x,y,p,q,a,g,k)\Leftrightarrow L(x,y,q^R,p^R,a,g,k).$$

The reversed values $p^R$ and $q^R$ depend on the properties of the attribute values $p$ and $q$ (e.g., scalar, vector, matrix) to be detailed in 8.5.

Of course, $\mathbf{P}_{\mathbf{RS}}$ is as well applicable to such an inference that 'if $x$ is to the right of $y$, then $y$ is to the left of $x$', which is conventionally based on a considerably large set of such *linguistic* axioms as (7-5) and (7-6). What is worse is that such formalization is not applicable to directions inexpressible by words such as 'left', 'under' and so on.

$$(\forall x,y)right(x,y)\supset left(y,x). \quad (7\text{-}5)$$

$$(\forall x,y)under(x,y)\supset above(y,x). \quad (7\text{-}6)$$

Consider Fig. 7-4 as for example of spatial change event of another attribute, namely, Height ($A_{03}$). This scene can be formulated as (7-7) and (7-8) whose expressions in natural language can be S7-16 and S7-17, respectively.

(S7-16) The building gets lower from east to west.

(S7-17) The building gets higher from west to east.

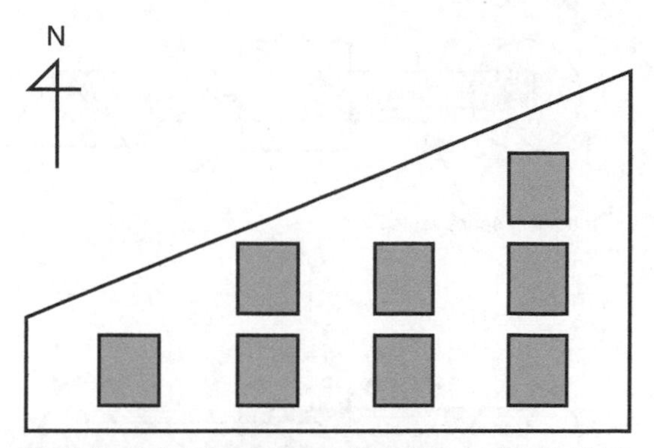


**Fig. 7-4.** Spatial change event of the attribute of height ($A_{03}$).

$$(\exists\ldots)\mathrm{L}(\mathrm{x,Building,p,q,}A_{03}\mathrm{,Gs,k})\Pi\mathrm{L}(\mathrm{x,Building,West,East,}A_{12}\mathrm{,Gs,k})\wedge \mathrm{p}<\mathrm{q} \quad (7\text{-}7)$$

$$(\exists\ldots)\mathrm{L}(\mathrm{x,Building,q,p,}A_{03}\mathrm{,Gs,k})\Pi\mathrm{L}(\mathrm{x,Building,East,West,}A_{12}\mathrm{,Gs,k})\wedge \mathrm{p}<\mathrm{q} \quad (7\text{-}8)$$

This is also the case for the pair of S7-18 and S7-19, formulated as (7-9) and (7-10), respectively, where ALT implies *altitude* as one of the standards of height.

(S7-18) The temperature of the air goes down as the altitude increases.
(S7-19) The temperature of the air goes up as the altitude decreases.

$$(\exists\ldots)\mathrm{L}(\mathrm{x,Air,}p_1,p_2,A_{03}\mathrm{,Gs,ALT})\wedge p_1>p_2 .\supset_0. (\exists\ldots)\mathrm{L}\ (\mathrm{z,Air,}q_1,q_2,A_{28}\mathrm{,Gs,k})\wedge q_1<q_2. \quad (7\text{-}9)$$

$$(\exists\ldots)\mathrm{L}(\mathrm{x,Air,}p_2,p_1,A_{03}\mathrm{,Gs,ALT})\wedge p_1>p_2 .\supset_0. (\exists\ldots)\mathrm{L}\ (\mathrm{z,Air,}q_2,q_1,A_{28}\mathrm{,Gs,k})\wedge q_1<q_2. \quad (7\text{-}10)$$

### *7.1.7 Arbitrary production of imaginary space region*

It is hypothesized that people can produce imaginary space regions arbitrarily in the attribute space of Physical location ($A_{12}$) within their mental image. *Postulate of Arbitrary Production of Imaginary Space Region* denoted as $\mathbf{P}_{\mathbf{IS}}$ is the formulation of this hypothesis as an unconditional postulate, where the event causer $y$ is the holder of the mental image containing the imaginary space region(x) denoted as *ISR(x)*.

$\mathbf{P}_{\mathbf{IS}}$. $(\exists\ldots)\mathrm{L}(\mathrm{y,x,p,p,}A_{12}\mathrm{,Gt,k})\wedge \mathrm{ISR(x)}$

### *7.1.8 Partiality of matter*

Any matter is assumed to consist of parts in a structure (i.e., spatial change event), which is generalized as *Postulate of Partiality of Matter* denoted as $\mathbf{P}_{\mathbf{PM}}$ here. The postulate $\mathbf{P}_{\mathbf{PM}}$ can read that $x_1$ is the conjoint of $x_2$ and $x_3$ (or that $x_2$ and $x_3$ conjoin into $x_1$). For example, Fig. 7-4 shows that an imaginary space region $x_1$ can be deemed as a complex of imaginary space regions $x_2$ and $x_3$.

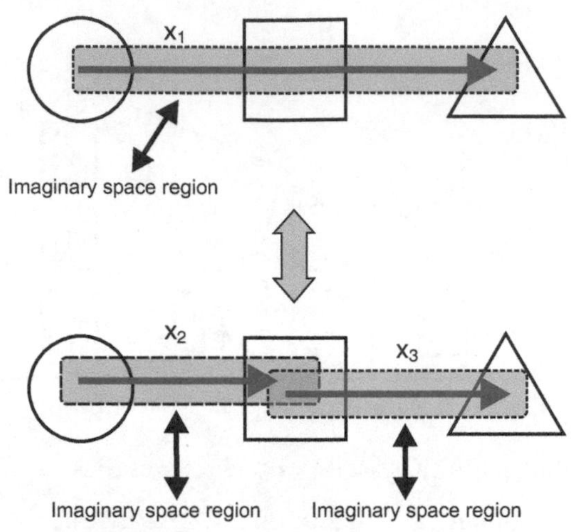


**Fig. 7-5.** Partiality of imaginary space region.

$\mathbf{P}_{PM}$. $(\forall\ldots)(\exists x_2, x_3)(L(y,x_1,p,q,a,G_s,k)\bullet L(y,x_1,q,r,a,G_s,k)$
$.\supset_0.L(y,x_2,p,q,a,G_s,k)\Pi L(y,x_3,q,r,a,G_s,k))$.
$(\forall\ldots)(\exists x_1)(L(y,x_2,p,q,a,G_s,k)\Pi L(y,x_3,q,r,a,G_s,k)$
$.\supset_0. L(y,x_1,p,q,a,G_s,k)\bullet L(y,x_1,q,r,a,G_s,k))$.

We refer to parts of an image especially for deductive inference upon it. For example, we can easily deduce from the top to the bottom in Fig. 7-5 such two facts that the square is to the left of the triangle and that the circle is to the left of the square. As its reversal, we can merge these two partial images into one meaningful image. That is, $\mathbf{P}_{PM}$ is very useful to compute static spatiotemporal (i.e., 4D) relations that are expressed by English spatial terms and conventionally formalized by a large set of such linguistic axioms as (7-11) and (7-12) below, as well as the case of the postulate $\mathbf{P}_{RS}$.

$$(\forall\ldots)\text{left}(y,x)\wedge\text{left}(z,y)\supset\text{between}(y,z,x). \quad (7\text{-}11)$$

$$(\forall\ldots)\text{under}(y,x)\wedge\text{under}(z,y)\supset\text{between}(y,z,x). \quad (7\text{-}12)$$

## 7.2 Inference Rules for Deduction

As easily understood from the definition of tempo-logical connectives, $\boldsymbol{L}_{md}$ expressions are wffs (well-formed formulas) of conventional predicate logic and, therefore, it is possible to apply any conventional inference rule to $\boldsymbol{L}_{md}$ expressions. This section introduces several examples of such inference rules involving tempo-logical connectives that have been actually implemented in our conversation management system to be described in Chapter 11. They are Commutativity Law of SAND, Elimination Law of SAND, Elimination Law of CAND, and

Substitution Law of locus formulas as follows. The validity of each inference rule is trivial from the definition of tempo-logical connectives.

$\mathbf{I}_{\mathbf{CS}}$. Commutativity Law of SAND:
$X \sqcap Y \leftrightarrow Y \sqcap X$

$\mathbf{I}_{\mathbf{ES}}$. Elimination Law of SAND:
$X \sqcap Y \rightarrow X$

$\mathbf{I}_{\mathbf{EC}}$. Elimination Law of CAND:
$X \bullet Y \rightarrow X, X \bullet Y \rightarrow Y$

$\mathbf{I}_{\mathbf{SL}}$. Substitution Law of locus formulas
$X(\alpha) \wedge \alpha \supset_0 \beta \rightarrow X(\beta)$,
where $\alpha$ and $\beta$ are arbitrary locus formulas.

## 7.3 Tempo-Logical Deduction with Tempo-logical Connectives

One of the most remarkable computations with tempo-logical connectives is **tempo-logical syllogism**, as is formalized by (7-13), where logical and temporal relations are calculated simultaneously in context of multiple tempo-logical implications. $\boldsymbol{\Gamma} \vdash A$ reads that A is deducible from hypothesis $\boldsymbol{\Gamma}$ (a set of assumptions) or that A is provable by $\boldsymbol{\Gamma}$.

$$P_1 \supset_i P_2, P_2 \supset_j P_3 \vdash P_1 \supset_k P_3, \tag{7-13}$$
where $\tau_i(P_1,P_2) \wedge \tau_j(P_2,P_3) .\supset. \tau_k(P_1,P_3)$.

Consider a sequence of temporal implications (7-14) as a proof of (7-15).

$$P \supset_{k1} X_1, X_1 \supset_{k2} X_2, \ldots, X_{m-1} \supset k_{km} Q \tag{7-14}$$

$$P \supset_n Q \tag{7-15}$$

For example, consider the propositions A-F below and we can understand F can be deduced from D and E.

A='Tom studies'
B='Tom is scolded'
C='Tom is given candies'
D='Tom does not study unless he is scolded in advance'
E='Tom studies immediately before he is given candies'
F='Tom is not given candies unless he is scolded in advance',

where D, E and F are formulated as (7-16)–(7-18), respectively.

$$D .\equiv. \sim B \supset_5 \sim A \tag{7-16}$$

$$E .\equiv. C \supset_{-1} A \tag{7-17}$$

$$F .\equiv. \sim B \supset_5 \sim C \tag{7-18}$$

The proof is as follows, where the theorem $\mathbf{T}_{\mathbf{TR51}}$ (or (5-14)) is inevitably employed.

$\mathbf{T}_{\mathbf{TR51}}$. $\tau_5(\chi_1,\chi_2)\wedge\tau_1(\chi_2,\chi_3) .\supset. \tau_5(\chi_1,\chi_3)$

(Proof)

E, $\mathbf{T}_{\mathbf{TC}}$ : $\sim A \supset_1 \sim C$ Con_1

D, Con_1 : $(\sim B \supset_5 \sim A) \wedge (\sim A \supset_1 \sim C)$

**PL,** $\mathbf{D}_{\mathbf{TL}}$, $\mathbf{T}_{\mathbf{TR51}}$ : $\rightarrow \sim B \supset_5 \sim C$

**[Q.E.D.]**

# 8

# Human-Specific Semantics of 4D Language as Mental Images

Natural language is the best communication means for ordinary people, which is also the case for casual interaction between ordinary people and home robots, so called here, intuitive human-robot interaction. Among all its sublanguages, spatiotemporal or 4D language has received particular attention in the field of ontology because its constituent concepts stand in highly complex relationships with underlying physical reality, accompanied with fundamental issues in terms of human cognition (for example, ambiguity, vagueness, temporality, identity, ...) appearing in varied subtle expressions (Harding, 2002). Conventionally, the term "spatial language" refers to the sublanguage of 4D (i.e., 3D space and time). On the other hand, spatial prepositions are termed exclusively for 3D space or, more exactly, 4D without change in time, namely, static 4D versus dynamic 4D. This chapter analyses human intuitive static 4D concepts as mental imagery and tries to formulate them as spatial change events in mental image description language (i.e., $\boldsymbol{L}_{md}$), leading to the 4D language understanding methodology for robots detailed in Chapter 11.

## 8.1 Conventional Approaches to 4D Language Semantics

Conventional approaches to spatial language semantics have inevitably employed a tremendously great number of axioms, such as (8-1), based on linguistic primitive concepts *without grounding in the real world* which is indispensable for robotic natural language understanding. It is noticeable that these axioms are part of the definition of 'between', valid only for verbalized directions such as 'left' and 'above', and that actually many more axioms should be provided for other directions such as 'before' and 'behind'. As easily understood, however, such a formulation is not so useful for artificial intelligence to understand 4D language, considering its coverage and efficiency.

$(\forall x,y)$ right(x,y) ≡ left(y,x)
$(\forall x,y)$ above(x,y) ≡ under(y,x)
$(\forall x,y,z)$ above(y,x) & above(x,z) ⊃between(x,y,z) (8-1)
$(\forall x,y,z)$ right(y,x) & right(x,z) ⊃between(x,y,z)
$(\forall x,y,z)$left(y,x) & left(z,y)⊃between(y,z,x).
.........

Moreover, viewed from cognitive robotics, this kind of formulation is inappropriate for a robot to act in the real world according to 4D expressions given by people. That is, it is too symbolic to ground 4D expressions in the world systematically through robotic sensation. On the contrary, mental image directed semantic theory gives the definition of 'between' in a simple and language-free formula, as in (8-5) given later in this chapter, which is applicable to every direction, whether or not in correspondence with words.

Overviewing conventional methodologies, almost all of them have provided robotic systems with such quasi-natural language expressions as 'bring (Agent, Recipient, Object)', 'move (Agent, Object, Source, Goal)', 'find (Agent, Object)', etc., for human instruction or suggestion, uniquely related to computer programs for deploying sensors/motors as their semantics (e.g., Coradeschi and Saffiotti, 2003; Drumwright et al., 2006; Wächtera et al., 2018). These expression schemata, however, are too linguistic or coarse to represent and compute sensory/motory events in such an integrative way as the intuitive human-robot interaction intended here. For example, consider S8-1, a command to a robot from a human, whose semantic interpretation could be given as (8-2) conventionally but he or she does not intend any specific direction expressible in word by 'between'.

(S8-1) Bring me the box between the table and the desk.

bring (Robot, Me, Box) & between(Box, Table, Desk). (8-2)

This is also the case for artificial intelligence planning ('action planning') which deals with the development of representation languages for planning problems and with the development of algorithms for plan construction (Wilkins and Myers 1995). However, conventional knowledge representation languages (i.e., KRLs) are yet too naïve to represent dynamic 4D events. For a serious example, it is noticeable that conventional AND denoted as *&* here is ambiguous in a temporal relationship. As for (8-2), the situation 'between(Box, Table, Desk)' does not necessarily hold after or during the event 'bring(Robot, Me, Box)'.

## 8.2 4D Language Semantics as Mental Images

Recently, location measurement systems like global positioning system have been widely utilized to obtain locations in the global coordinates system and have already been applied to location-aware systems of practical use, such as vehicle-navigation systems. It is, however, not very common for people to refer to such pinpoints, rather to intuitive static 4D relations such as *at, around, along*, etc., in their casual communication. Therefore, in order to facilitate intuitive human-robot interaction, 4D language is the most important of all sublanguages, especially when both the entities must share knowledge of 4D arrangement of home utilities such as desk, table, etc.

From the semantic viewpoint, 4D expressions have the virtue of relating in some way to visual scenes being described. Actually, we can have a certain mental image evoked by a 4D expression as its plausible interpretation, just like a movie. To the author's best knowledge, most conventional approaches to 4D language understanding have focused on computing purely objective 3D (more exactly, static 4D) geometric relations (i.e., topological, directional and metric relations) conceptualized as spatial prepositions or so, sometimes considering properties and functions of the objects involved (e.g., Logan and Sadler, 1996; Coventry, Prat-Sala and Richards, 2001) but discarding such verb-centered 3D expressions as S8-2–S8-11.

(S8-2) The path *climbs* zigzag up the slope.

(S8-3) The river *zigzags* through the countryside.

(S8-4) The pavement *reaches* the center of the city.

(S8-5) The skyscraper *towers* high to the heavens.

(S8-6) The tower *protrudes* above its built environment.

(S8-7) The path *rises* from the seashore to the hill.

(S8-8) The path *sinks* to the seashore from the hill.

(S8-9) The two paths *meet* at the point.

(S8-10) The two paths *separate* at the point.

(S8-11) The mountain chain extends to the north.

These verb concepts are considered to reflect human cognitive activities toward the physical world. Therefore, they can be subjective to each human individual but can also be common to the human beings in conceptualization of their environment. Mental image directed semantic theory can provide robots with the capacity to understand 4D language in a human-like way based on the formal system already introduced in Chapter 6, where human cognitive propensities toward the external world as human-likeness are given by a set of postulates.

In mental image directed semantic theory, dynamic 4D and static 4D events respectively correspond with temporal change events and spatial change events in Physical location ($A_{12}$), discriminated by the pattern parameter in $\boldsymbol{L}_{md}$ (i.e., $g$=$G_t$ or $G_s$). The former are conceptualized by such verbs as *carry*, *return* and *fetch* whose formulations with $g$=$G_t$ are presented in Chapter 6. On the other hand, the latter are mainly verbalized as spatial prepositions or such verbs as those in S8-2–S8-11, to be formulated with $g$=$G_s$.

An event expressed in $\boldsymbol{L_{md}}$ is compared to a movie film taken through a floating camera where both temporal and spatial extensions of the event are recorded as a time sequence of snapshots because it is necessarily grounded in the movements of the focus of attention of the observer over the event. This is the reason why $\boldsymbol{L_{md}}$ can formalize both types of events in a systematically unified way and is one of the most remarkable features of $\boldsymbol{L_{md}}$, clearly distinguished from other KRLs.

There are two major hypotheses assumed on mental image. One is that mental image is in one-to-one correspondence with movement of the focus of attention of the observer, as mentioned above. The other is that it is not one-to-one reflection of the real world. It is well known that people perceive more than reality, for example, imaginary objects such as gestalt. Mental image directed semantic theory employs an imaginary object called imaginary space region (conventionally abbreviated as ISR) in order to formalize mental image systematically. For example, Fig. 8-1 concerns the perception of the formation of multiple objects, where the focus of attention of the observer runs along an imaginary space region surrounded by a dotted line. This spatial event (i.e., $g=G_s$) can be verbalized as S8-12–S8-14 using the preposition *between* and formulated as (8-3) or (8-4), corresponding also to such concepts as *row*, *line up*, etc., In these formulas, the constants $A_{13}$, →, and ← are for the attribute *Direction*, *rightward* and *leftward*, respectively. Moreover, the standard constant $K_o$ denotes that the direction values are given as unit vectors originated (or viewed) from the observer, for example, rightward as (0,1,0) and leftward as (0,-1,0) (see Fig. 8-5).

(S8-12) The square is between the triangle and the circle.

(S8-13) The circle, square, and triangle are lined up.

(S8-14) The circle, square, and triangle are in row.

$$(\exists\ldots)(L(z,y,x_1,x_2,A_{12},G_s,k)\bullet L(z,y,x_2,x_3,A_{12},G_s,k))\Pi$$
$$L(z,y,\rightarrow,\rightarrow,A_{13},G_s,K_o)\wedge ISR(y)\wedge circle(x_1)\wedge square(x_2)\wedge triangle(x_3). \quad (8\text{-}3)$$

$$(\exists\ldots)(L(z,y,x_3,x_2,A_{12},G_s,k)\bullet L(z,y,x_2,x_1,A_{12},G_s,k))\Pi$$
$$L(z,y,\leftarrow,\leftarrow,A_{13},G_s,K_o)\wedge ISR(y)\wedge circle(x_1)\wedge square(x_2)\wedge triangle(x_3). \quad (8\text{-}4)$$

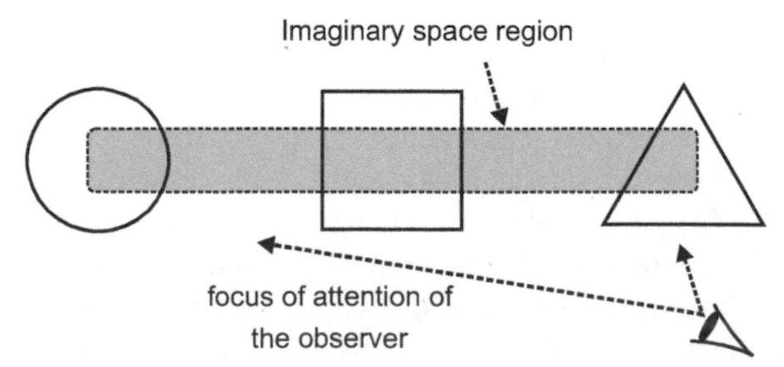


**Fig. 8-1.** Multiple objects and movement of the focus of attention of the observer.

By the way, the conventional definition of '$(x_1)$ between $(x_2$ and $x_3)$' can be given simply as (8-5) in $\boldsymbol{L}_{md}$, where the value $r$ of direction $(A_{13})$ implies the three objects should be in line, although this is not always the case, as pointed out later in this chapter.

$$(\lambda x_1,x_2,x_3)\text{between}(x_1,x_2,x_3)$$
$$\Leftrightarrow(\lambda x_1,x_2,x_3)(\exists\ldots)(L(z,y,x_2,x_1,A_{12},G_s,k_1)\bullet L(z,y,x_1,x_3,A_{12},G_s,k_1)) \quad (8\text{-}5)$$
$$\Pi L(z,y,r,r,A_{13},G_s,k_2)\wedge ISR(y).$$

With a special attention, the author has analyzed a considerable number of 4D terms over various kinds of English words, including prepositions, verbs, adverbs, etc., categorized as *Dimensions*, *Form* and *Motion* in the class *SPACE* of the Roget's thesaurus (Roget, 1975), and found that almost all the concepts of spatial change events can be defined in exclusive use of five kinds of attributes for the focus of attention of the observers, namely, Physical location $(A_{12})$, Direction $(A_{13})$, Trajectory $(A_{15})$, Mileage $(A_{17})$ and Topology $(A_{44})$. The values for each of these attributes are given in the forms as follows.

a) Physical location $(A_{12})$: Intuitive areas denoted by words such as *about* and *at* or scientific areas such as a set of global coordinates, containing or occupied by the attribute carrier

b) Direction $(A_{13})$: Intuitive values denoted by such words as *westward* and *leftward* or scientific 3D unit vectors such as (1,0,0), with origins indicated at the parameter Standard

c) Trajectory $(A_{15})$: Intuitive or scientific shapes denoted by such words as *point*, *line*, *circle*, *zigzag*, etc.

d) Mileage $(A_{17})$: Intuitive distances such as *far* and *near* or rigid values indicated by scientific units such as *meter* and *mile*

e) Topology $(A_{44})$: Intuitive values denoted by such words or phrases as *Disjoint* and *Covered-by*, graphically interpreted as Fig. 8-2 and scientifically given by 9-intersection matrices (Egenhofer, 1991) indicated by the Standard value $K_{9IM}$.

For example, S8-15 and S8-16 are semantically translated into (8-6) and (8-7), respectively, where *In*, *Cont* and *Dis* are the values *inside*, *contains* and *disjoint*, respectively. Practically, these topological values are given as 3×3 matrices with each element equal to 0 or 1 and, therefore, for example, *In* and *Cont* are transposes of each other.

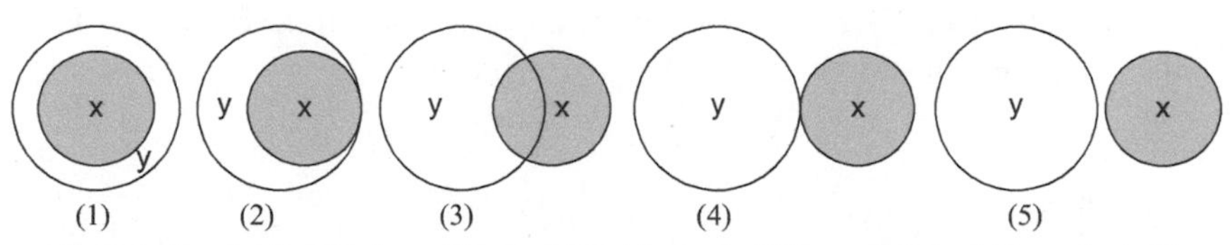


**Fig. 8-2.** Topological relations: (1)In(x,y)/Contains(y,x), (2)Covered_by(x,y)/Covers(y,x), (3) Overlaps(x,y)/Overlaps(y,x), (4) Meets(x,y)/Meets(y,x), (5) Disjoint(x,y)/Disjoint(y,x).

(S8-15) Tom is in the room.

$(\exists\ldots)L(Tom,x,y,Tom,A_{12},G_s,k)\ \Pi L(Tom,x,In,In,A_{44},G_t,K_{9IM}) \wedge ISR(x) \wedge room(y).$

or (8-6)

$(\exists\ldots)L(Tom,x,Tom,y,A_{12},G_s,k)\Pi\ L(Tom,x,Cont,Cont,A_{44},G_t,K_{9IM}) \wedge ISR(x) \wedge room(y).$

(S8-16) Tom exits the room.

$(\exists\ldots)L(Tom,Tom,p,q,A_{12},G_t,k_1)\Pi L(Tom,x,y,Tom,A_{12},G_s,k_1)$ (8-7)

$\Pi L(Tom,x,In,Dis,A_{44},G_t,K_{9IM})$

$\wedge ISR(x) \wedge room(y) \wedge p \neq q.$

or

$(\exists\ldots)L(Tom,Tom,p,q,A_{12},G_t,k_1)\Pi L(Tom,x,Tom,y,A_{12},G_s,k_1)$

$\Pi L(Tom,x,Cont,Dis,A_{44},G_t,K_{9IM})$

$\wedge ISR(x) \wedge room(y) \wedge p \neq q.$

## 8.3 Formulation of Concepts of Spatial Prepositions

The concepts of the 41 English spatial prepositions listed below were analyzed and formulated in accordance with mental image directed semantic theory. To be most remarkable, the concepts of spatial prepositions are defined as 4D images in mental image directed semantic theory but not as 3D (=4D minus 'time') images in conventional approaches.

{about, in, at, to, into, inside, within, out of, outside, from, through, off, on, above, over, up, under, underneath, below, down, beneath, between, among, amid, round, around, along, across, toward, for, before, in front of, after, behind, by, via, beside, near, against, beyond, next}

Hereafter, consider every spatial preposition *P* to appear in either of the two contexts in (8-8) where *V* denotes a verb and assume its concept to be denoted as *P*(*x*,*y*), representing the spatial relation of two objects *x* and *y,* including 'Between (*x*,*y*,*z*)' as a specialcase. Most of the spatial prepositions are used both for temporal and spatial change events, where whether g=$G_t$ or $G_s$ is to be determined by the semantic contexts about *x* or *V* in each case. For simplicity, the existential quantifiers (i.e., ∃) are to be omitted in the definitions below. Needless to say, all the definitions here are subjective to the author, which, if necessary, are to be denoted by a certain standard value at $\omega_7$ in (6-1) explicitly.

x V P y/x P y. (8-8)

A) At/About

According to Webster's, *at* is used as a function word to indicate presence or occurrence *in*, *on*, or *near* something, leading to such a subjective image as

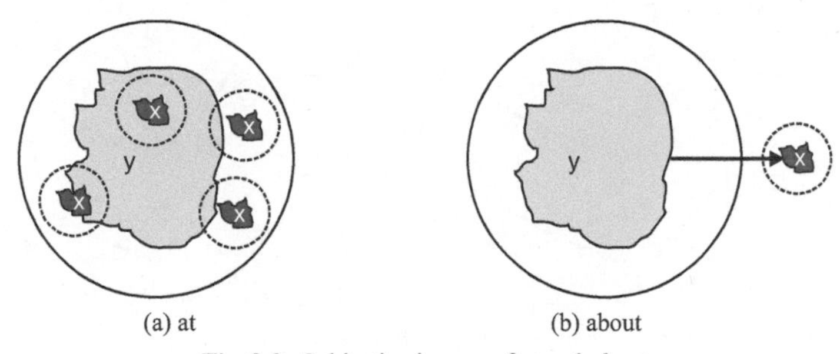


**Fig. 8-3.** Subjective images of *at* and *about.*

Fig. 8-3a. In $\boldsymbol{L}_{md}$, this concept is defined as simply as (8-9), reading that object *x* is included in a certain subjective region (depicted as a circle here) surrounding object *y* that does not necessarily match the region physically occupied by *y*.

$$(\lambda x,y)at(x,y)\Leftrightarrow(\lambda x,y)(\exists\ldots)L(z,x,y,y,A_{12},G_t,k). \quad (8\text{-}9)$$

On the other hand, *about* is defined as '*x* in the immediate neighborhood of *y*' in Webster's and as (8-10) in $\boldsymbol{L}_{md}$, associating with the image in Fig. 8-3b. The value $D_n$ represents a certain subjective maximum mileage for *x* and *y* to neighbor each other. The mileage range *[0,m]* implies that the focus of attention runs upto *m* (kilo-meter or so).

$$(\lambda x,y)about(x,y)\Leftrightarrow(\lambda x,y)(\exists\ldots)L(z_1,z,y,x,A_{12},Gs,k_1)\Pi L(z_2,z,0,m,A_{17},Gs,k_2)\wedge 0<m \leq D_n\wedge ISR(z). \quad (8\text{-}10)$$

B) Along

Webster's defines *along* as 'in a line matching the length or direction of something'. Mental image directed semantic theory, however, adopts (8-11) as its definition concerning the direction $A_{13}$ only, taking account of such examples as S8-17–S8-21. This formula implies that the direction of *x* is the same as *y* temporally or spatially. There are four possibilities as shown in Fig. 8-4, where each solid curve stands for an object's spatial extension and each broken curved arrow for an object's movement.

$$(\lambda x,y)along(x,y)\Leftrightarrow (\lambda x,y)(\exists\ldots)L(z,x,y,y,A_{13},g,k). \quad (8\text{-}11)$$

(S8-17) Tom is walking along the road.

(S8-18) The building is along the river.

(S8-19) Tom is swimming along (with) the flow.

(S8-20) The road runs along the coast.

(S8-21) Many people stood along the marching.

C) Across/Above/Underneath

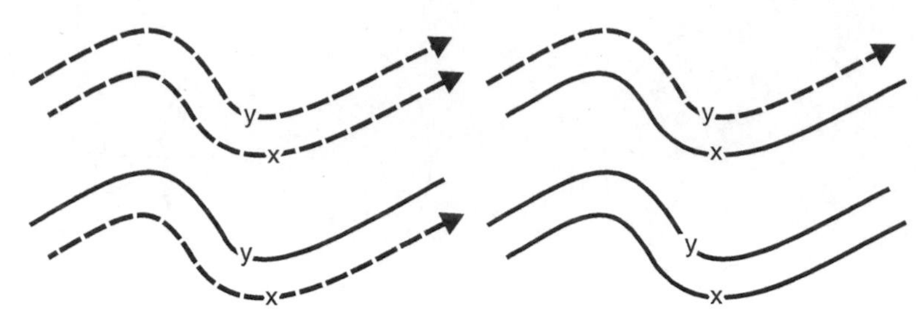


**Fig. 8-4.** Four possibilities of *along*.

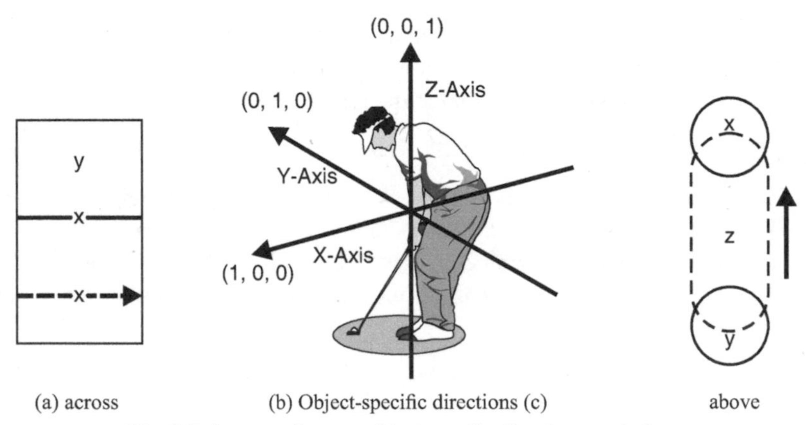


(a) across (b) Object-specific directions (c) above

**Fig. 8-5.** Images of *across*, object-specific directions, and *above.*

The images of 'across' is shown in Fig. 8-5a with two possibilities to be formulated as (8-12), where the value *d* denotes 'lateral direction' of an object (*y*) as shown in Fig. 8-5b. Figure 8-5c and (8-13) are for 'above' and (8-14) is for 'underneath'.

$$(\lambda x,y)\text{across}(x,y)\Leftrightarrow(\lambda x,y)(\exists\ldots)L(z_1,x,d,d,A_{13},g,y)\Pi L(z_2,z,y,x,A_{12},Gs,k)\Pi (L(z_3,z,\text{Covered\_by},\text{In},A_{44},g,K_{9IM})\bullet L(z_4,z,\text{In},\text{Covered\_by},A_{44},g,K_{9IM}))\wedge d=(0,\pm1,0)\wedge ISR(z). \quad (8\text{-}12)$$

$$(\lambda x,y)\text{above}(x,y)\Leftrightarrow(\lambda x,y)(\exists\ldots)L(z_1,z,y,x,A_{12},G_s,k_1)\Pi L(z_2,z,\uparrow,\uparrow,A_{13},G_s,k_2)\Pi L(z_3,z,\text{Dis},\text{Dis},A_{44},G_t,K_{9IM})\wedge ISR(z). \quad (8\text{-}13)$$

$$(\lambda x,y)\text{underneath}(x,y)\Leftrightarrow(\lambda x,y)(\exists\ldots)L(z_1,z,y,x,A_{12},G_s,k_1)\Pi L(z_2,z,\downarrow,\downarrow,A_{13},G_s,k_2)\Pi L(z_3,z,\text{Meet},\text{Meet},A_{44},G_t,K_{9IM})\wedge ISR(z). \quad (8\text{-}14)$$

D) Between

Conventionally, as mentioned above, the semantic definition of '($x_1$) *between* ($x_2$) and ($x_3$)' is given by (8-5) necessarily referring to the direction ($A_{13}$). This definition is valid only for such a scene as shown in Fig. 8-1 where an imaginary object *z* (i.e., imaginary space region) lies over three real objects $x_1$, $x_2$, and $x_3$,

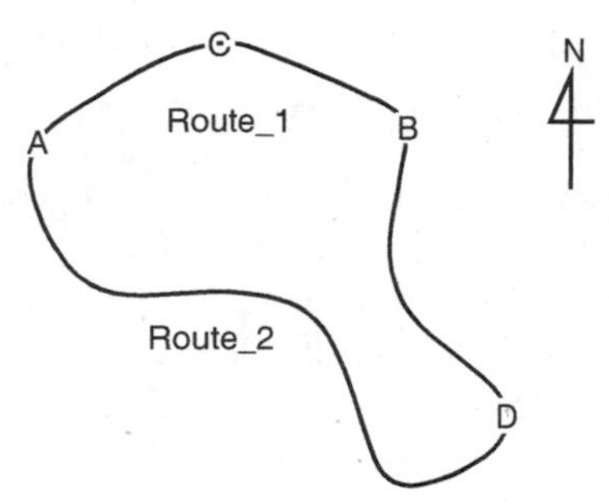


**Fig. 8-6.** Map: Towns and Routes.

*in line*, namely, of a good *gestalt quality*. However, there is yet another case where the direction is indifferent to the definition. It is noticeable that various possibilities about one spatial relation with different reference frames can be treated in mental image directed semantic theory. Consider the map shown in Fig. 8-6 whose $\boldsymbol{L}_{md}$ expression is given as (8-15).

$$(L(\_,Route\_1,A,C,A_{12},G_s,\_)\bullet L(\_,Route\_1,C,B,A_{12},G_s,\_))\Pi$$
$$(L(\_,Route\_2,A,D,A_{12},G_s,\_)\bullet L(\_,Route\_2,D,B,A_{12},G_s,\_)) \quad (8\text{-}15)$$

About this map, people could answer the question S8-22 by S8-23 or S8-24, each of which describes three real objects $x_1$, $x_2$, and $x_3$, (i.e., towns) located on a curvilinear object (i.e., route) as a reference frame.

(S8-22) What town is between town A and B?

(S8-23) Town C is between town A and B (on Route_1).

(S8-24) Town D is between town A and B (on Route_2).

Not to mention, such a conventional definition such as (8-1) or its counterpart (8-5) in $\boldsymbol{L}_{md}$ cannot treat this kind of problem. For Route_1, such a predicate as 'west' in (8-16) is employable, but it is not the case for Route_2. Within conventional natural language understanding, another predicate such as 'link' in (8-17) should be devised for Route_2 that is applicable to Route_1 as well. This kind of enforcement, however, is too *ad hoc* to establish a systematic framework for natural semantics.

$$west(A,C)\ \&\ west(C,B)\supset between(C,A,B) \quad (8\text{-}16)$$

$$link(A,D)\ \&\ link(D,B)\supset between(D,A,B) \quad (8\text{-}17)$$

In this case, another definition of between is simply given by (8-18) where the standard parameter *k* is employed in order to denote each reference frame. That is, S8-23 and S8-24 are interpreted as (8-19) and (8-20), respectively.

$$(\lambda x_1,x_2,x_3)between(x_1,x_2,x_3) \quad (8\text{-}18)$$
$$\Leftrightarrow(\lambda x_1,x_2,x_3)(\exists\ldots)L(y,z,x_2,x_1,A_{12},G_s,k)\bullet L(y,z,x_1,x_3,A_{12},G_s,k)\wedge ISR(y).$$

$(\exists\ldots)L(\_,y,A,C,A_{12},G_s,\text{Route\_1})\bullet L(\_,y,C,B,A_{12},G_s,\text{Route\_1})\wedge ISR(y)$ (8-19)

$(\exists\ldots)L(\_,y,A,D,A_{12},G_s,\text{Route\_2})\bullet L(\_,y,D,B,A_{12},G_s,\text{Route\_2})\ ISR(y)$ (8-20)

This definition does not refer to the direction ($A_{13}$) of the focus of attention of the observer running through the three objects but the reference frame is to be put explicitly at the standard *k* in its actual adaption. That is, the direction of its movement is not always mandatory for the definition of 'between' but its visiting order denoted in the locus formula.

E) In/On

According to Webster's, *at* is used as a function word to indicate presence or occurrence *in*, *on*, or *near* something, which means that *in* $(x,y)$ and *on* $(x,y)$ imply *at* $(x,y)$. Therefore, the concepts of *in* and *on* can be defined as (8-21) and (8-22), respectively.

$(\lambda x,y)in(x,y)\Leftrightarrow(\lambda x,y)(\exists\ldots)L(z_1,z,y,x,A_{12},G_s,k)\Pi LL(z_2,z,In,In,A_{44},G_t,K_{9IM})\wedge ISR(z)$ (8-21)

$(\lambda x,y)on(x,y)\Leftrightarrow(\lambda x,y)(\exists\ldots)L(z_1,z,y,x,A_{12},G_s,k)\Pi L(z_2,z,Meet,Meet,A_{44},G_t,K_{9IM})\wedge ISR(z).$ (8-22)

The alternative definition (8-23) can be better for *on* used as a function word to indicate position in contact with and supported by the top surface of something.

$(\lambda x,y)on(x,y)\Leftrightarrow(\lambda x,y)(\exists\ldots)L(z_1,z,y,x,A_{12},G_s,k)\Pi L(z_2,z,Meet,Meet,A_{44},G_t,K_{9IM})\Pi L(z_3,z,\uparrow,\uparrow,A_{13},G_s,k2)\wedge ISR(z).$ (8-23)

It is very natural to introduce such a postulate as $\mathbf{P}_{at}$ (*Postulate of preposition at*) as follows. This is one of the cases where attributes extracted from natural concepts are not necessarily independent.

$\mathbf{P}_{at}$ $(\forall\ldots)L(z_1,z,y,x,A_{12},G_s,k)\Pi LL(z_2,z,Tat,Tat,A_{44},G_t,K_{9IM})\wedge ISR(z).\supset_0. L(z_3,x,y,y,A_{12},G_t,k),$

where *Tat=In*, *Covered_by*, *Overlaps* or *Meet*.

## 8.4 Properties of Static 4D Concepts as Human Intuitive Mental Images

Here, $\boldsymbol{L}_{md}$ is applied in order to formalize complicated spatial scenes and demonstrate its descriptive power.

### *8.4.1 Complexity of static 4D concepts*

The mathematically rigid topology between two objects as shown in Fig. 8-2 must be determined with the perfect knowledge of their insides, outsides and boundaries

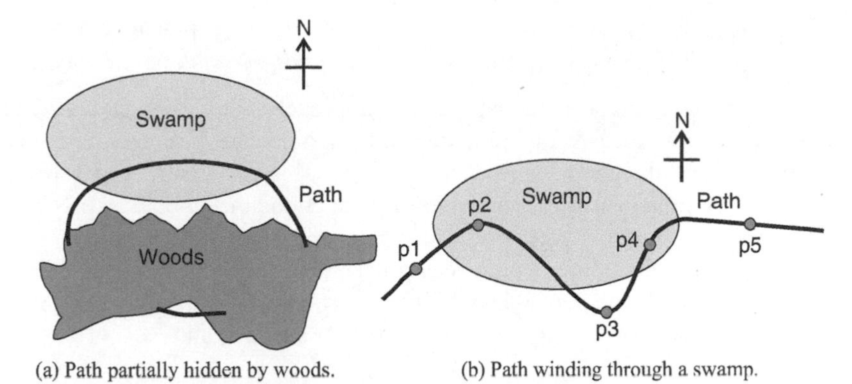


(a) Path partially hidden by woods. (b) Path winding through a swamp.

**Fig. 8-7.** Delicate topological relations.

(Egenhofer, 1991; Shariff et al., 1998). Ordinary people, however, would often comment on matters without knowing all about them. This is the very case when they encounter an unknown object too large to observe at a glance, like a road in a strange country. For example, Fig. 8-7a shows such a path viewed from the sky that is partly hidden in the woods. In this case, the topological relation between the path as a whole and the swamp/woods depends on how the path starts and ends in the woods, but people could utter such sentences as S8-25 and S8-26 about this scene.

(S8-25) The path goes into the swamp/woods.

(S8-26) The path comes out of the swamp/woods.

Actually, these sentences refer to such events that reflect certain temporal changes in the topological relation between the swamp/woods and the focus of attention of the observer running along the path. Therefore, their conceptual descriptions are to be given as (8-24) and (8-25), respectively.

$$(\exists\ldots)L(\_,z,p,q,A_{12},G_s,\_)\Pi L(\_,x,y,z,A_{12},G_s,\_)\Pi L(\_,x,Dis,In,A_{44},G_s,K_{9IM})\wedge ISR(x) \wedge\{swamp(y)/woods(y)\}\wedge path(z) \wedge p\neq q. \quad (8\text{-}24)$$

$$(\exists\ldots)L(\_,z,p,q,A_{12},G_s,\_)\Pi L(\_,x,y,z,A_{12},G_s,\_) \quad (8\text{-}25)$$
$$\Pi\ L(\_,x,In,Dis,A_{44},G_s,K_{9IM})\wedge ISR(x)\wedge\{swamp(y)/woods(y)\}\wedge path(z) \wedge p\neq q.$$

For another example, Fig. 8-7b shows a more complicated spatial change event in topology that can be formulated as (8-26) and could be verbalized as S8-27.

$$(\exists\ldots)L(\_,z,y,x,A_{12},G_s,\_)\Pi((L(\_,x,p_1,p_2,A_{12},G_s,\_)\Pi\ (L(\_,z,Dis,In,A_{44},G_s,K_{9IM})) \quad (8\text{-}26)$$
$$\bullet(L(\_,x,p_2,p_3,A_{12},G_s,\_)\Pi L(\_,z,In,Dis,A_{44},G_s,K_{9IM}))\bullet\ (L(\_,x,p_3,p_4,A_{12},G_s,\_)$$
$$\Pi L(\_,z,Dis,In,A_{44},G_s,K_{9IM}))\ \bullet(L(\_,x,p_4,p_5,A_{12},G_s,\_)\Pi\ L(\_,z,In,Dis,A_{44},G_s,K_{9IM})))$$
$$\wedge path(x)\wedge swamp(y)\wedge ISR(z).$$

(S8-27) The path cuts the swamp twice as shown in Fig. 8-7b, passing $p_1$ outside, $p_2$ inside, $p_3$ outside, $p_4$ inside and $p_5$ outside the swamp on the way.

As already mentioned above, most approaches to spatial language understanding have focused on computing purely objective geometric relations (i.e., topological, directional and metric relations) conceptualized as spatial prepositions or so, considering properties and functions of the objects involved (Logan et al., 1996; Coventry et al., 2001; Egenhofer, 1991). Although these kinds of geometric relations are not necessarily independent of one another, here topological relation is focused on in order to demonstrate the descriptive power of $\boldsymbol{L}_{md}$ for spatial change events that conventional KRLs lack at all.

### 8.4.2 Paraphrasing of mental image description language expressions by semantic re-articulation

As mentioned above, most 4D verbs can be employed for spatial scenes which are too complicated for spatial prepositions to express. As also mentioned above, all loci in attribute spaces are assumed to correspond one to one with movements or, more generally, temporal change events of the focus of attention of the observer, which in turn affects semantic articulation of sensation resulting in different $\boldsymbol{L}_{md}$ expressions. This assumption is introduced to the formal system as Postulate of arbitrariness in locus articulation-type1 (i.e., $\mathbf{P}_{A1}$) and Postulate of arbitrariness in locus articulation-type2 (i.e., $\mathbf{P}_{A2}$). In particular, the $\boldsymbol{L}_{md}$ expression of a spatial change event is quite subjective to the involved movement of the focus of attention of the observer (or reflects the movement itself). This fact implies that representation in $\boldsymbol{L}_{md}$ does not necessarily guarantee semantic normalization of various natural language verbalizations of a scene, or more precisely, its sensation. Postulate of Reversibility of Spatial Change Event (i.e., $\mathbf{P}_{RS}$) and Postulate of Partiality of Matter (i.e., $\mathbf{P}_{PM}$) are introduced in order to enrich paraphrasing among observer-oriented $\boldsymbol{L}_{md}$ expressions of the same static 4D scene by semantic re-articulation alongside $\mathbf{P}_{A1}$ and $\mathbf{P}_{A2}$.

For example, consider S8-28, 29, and 30 which describe the same scene where the Andes mountains are located in a line. Their interpretations in $\boldsymbol{L}_{md}$ differ as given by (8-27), (8-28), and (8-29), respectively. Most event causers in spatial change events are unknowable or of no significance for further reasoning based on their conceptual definitions. Hence, the anonymous variable '_' is often employed for easier reading.

(S8-28) The Andes mountains run north.

(S8-29) The Andes mountains run south.

(S8-30) The Andes mountains run north and south.

$$(\exists\ldots)L(\_,Andes,p,q,A_{12},G_s,\_)\Pi L(\_,Andes,Nth,Nth,A_{13},G_s,\_)\wedge p\neq q. \qquad (8\text{-}27)$$

$(\exists\ldots)L(\_,Andes,p,q,A_{12},G_s,\_)\Pi L(\_,Andes,Sth,Sth,A_{13},G_s,\_)\wedge p\neq q.$ (8-28)

$(\exists\ldots)L(\_,Andes,p,q,A_{12},G_s,\_)\Pi L(\_,Andes,Nth,Nth,A_{13},G_s,\_)\Pi$ (8-29)
$L(\_,Andes,Sth,Sth,A_{13},G_s,\_)\wedge p\neq q.$

As easily understood, (8-27) and (8-28) can be proved as equivalent each other simply by adapting the postulate $\mathbf{P}_{\mathbf{RS}}$. On the other hand, (8-29) can be reduced to (8-27) or (8-28) in use of $\mathbf{P}_{\mathbf{RS}}$ followed by the inference rule *Elimination Law of SAND*.

For another example, S8-31, 32, and 33 describe one of the sub-structures of the human body. The first two sentences are to be directly translated into $\boldsymbol{L}_{md}$ expressions (8-30) and (8-31), respectively. On the other hand, the semantic interpretation of S8-33 into (8-32) requires application of postulate of partiality of matter and postulate of reversibility of spatial change event to (8-30) and (8-31). Finally, in order to affirm the question S8-34, (8-33) is to be deduced from (8-32) in use of postulate of arbitrariness in locus articulation-type1.

(S8-31) The upper arm is up from the elbow to the shoulder.

(S8-32) The lower arm is down from the elbow to the wrist.

(S8-33) The arm is the conjoint of the lower and the upper arms.

(S8-34) Is the arm down from the shoulder to the wrist?

$L(\_,U\text{-}arm,El,Sh,A_{12},G_s,\_)\Pi L(\_,U\text{-}arm,\uparrow,\uparrow,A_{13},G_s,\_).$ (8-30)

$L(\_,L\text{-}arm,El,Wr,A_{12},G_s,\_)\Pi L(\_,U\text{-}arm,\downarrow,\downarrow,A_{13},G_s,\_).$ (8-31)

$(L(\_,Arm,Sh,El,A_{12},G_s,\_)\bullet L(\_,Arm,El,Wr,A_{12},G_s,\_))\Pi L(\_,Arm,\downarrow,\downarrow,A_{13},G_s,\_).$
(8-32)

$L(\_,Arm,Sh,Wr,A_{12},G_s,\_)\Pi L(\_,Arm,\downarrow,\downarrow,A_{13},G_s,\_).$ (8-33)

## 8.5 Reversal Operation on Spatial Change Event Concepts as Mental Images

People think in mental image by operating or transforming it in various ways, such as mental rotation. Computations on formulated mental images in the formal system are performed as logical inferences based on the axioms for predicate logic and about 10 postulates for mental imagery operation specific to mental image directed semantic theory. For example, the postulate $\mathbf{P}_{\mathbf{RS}}$ is one of the principal postulates belonging to people's common-sense knowledge about geography. This postulate is accompanied with the mental operation *reversal*, denoted by *R* and recursively defined as (8-34), namely, Definition of Reversal Operation of Mental Image ($\mathbf{D}_{\mathbf{RO}}$), where $\chi_i$ stands for a locus.

$\mathbf{D}_{\mathbf{RO.}}$

$$(\chi_1 \bullet \chi_2)^R \Leftrightarrow \chi_2^R \bullet \chi_1^R$$
$$(\chi_1 \Pi \chi_2)^R \Leftrightarrow \chi_1^R \Pi \chi_2^R$$
$$L^R(x,y,p,q,a,g,k) \Leftrightarrow L(x,y,q^R,p^R,a,g,k). \quad (8\text{-}34)$$

The reversed values $p^R$ and $q^R$ depend on the properties of the attribute values $p$ and $q$ (e.g., scalar, vector, matrix). For example,

$p^R = p$, $q^R = q$ for Physical location ($A_{12}$) because $p$ and $q$ are given as (sets of) coordinates;

$p^R = q$, $q^R = p$ for Trajectory ($A_{15}$) because any trajectory begins with a point as $p$ and ends in its completion as $q$;

$p^R = q$, $q^R = p$ for Mileage ($A_{17}$) because mileage increases in time;

$p^R = -p$, $q^R = -q$ for Direction ($A_{13}$) because the values are given as vectors;

$p^R = p^T$, $q^R = q^T$ for Topology ($A_{44}$), where $T$ is the transpose operator for matrices.

The postulate $\mathbf{P}_{\mathbf{RS}}$ is formulated as (8-35), where $\chi_s$ and $\chi_s^R$ are an image and its *reversal* for a certain spatial change event, respectively. These two images are substitutable with each other because of the property of $\equiv_0$, that is, the inference rule Substitution Law of Locus Formulas (i.e., $I_{SL}$) is bi-directionally applicable. For example, when $\mathbf{P}_{\mathbf{RS}}$ is applied to (8-36), the semantic interpretation of S8-35, it is transformed into (8-37), its equivalent as the semantic interpretation of S8-36.

$$\chi_s^R . \equiv_{0.} \chi_s . \quad (8\text{-}35)$$

(S8-35) The railroad runs east 50 kms straight from A to B via C.

(S8-36) The railroad runs west 50 kms straight from B to A via C.

$$(\exists \ldots)(L(\_,x,A,C,A_{12},G_s,\_) \bullet L(\_,x,C,B,A_{12},G_s,\_)) \Pi L(\_,x,0,50\text{ km},A_{17},G_s,\_) \Pi$$
$$L(\_,x,Point,Line,A_{15},G_s,\_) \Pi L(\_,x,East,East,A_{13},G_s,\_) \wedge railroad(x). \quad (8\text{-}36)$$

$$(\exists \ldots)(L(\_,x,B,C,A_{12},G_s,\_) \bullet L(\_,x,C,A,A_{12},G_s,\_)) \Pi L(\_,x,0,50\text{ km},A_{17},G_s,\_) \Pi$$
$$L(\_,x,Point,Line,A_{15},G_s,\_) \Pi L(\_,x,West,West,A_{13},G_s,\_) \wedge railroad(x). \quad (8\text{-}37)$$

# 9 Problem Finding and Solving in Formal System

Imagine a scenario in which a robot is working in order to achieve the mission assigned to it by some people. The robot must find and solve problems concerning the mission. The problems here are considered to belong to the category *Exploration problem* introduced by Heylighen (Heylighen, 1988) for robots partially or wholly ignorant of their environments. Such problems can roughly be classified into two subcategories, as follows.

Creation Problem: House building, food cooking, etc.

Maintenance Problem: Fire extinguishing, room cleaning, etc.

In general, a maintenance problem is a relatively simple one that the robot can find and solve autonomously, while a creation problem is a relatively difficult one that is given to the robot, possibly by humans, to be solved in cooperation with them.

## 9.1 Definition of Problem and Task

The robot must determine its task to solve a problem in the world. In general, during such problem solving, the robot needs to interpolate some transit event $X_T$ between the two events, namely, *Current Event* ($X_C$) and *Goal Event* ($X_G$) as shown by (9-1).

$$X_C \bullet X_T \bullet X_G \quad (9\text{-}1)$$

According to this formalization, a problem $X_P$ is defined as $X_T \bullet X_G$ and a task for the robot is defined as its realization in the same way as the conventional artificial intelligence referred to by Russell and Norvig (Russell and Norvig, 2010), etc., where a problem is defined as the difference or gap between a *Current*

*State* and a *Goal State* and a task as its cancellation. Here, the term *Event* is preferred to the term *State*, and instead *State* is defined as static *Event* which corresponds to a level locus. The events in the world are described as loci in certain attribute spaces and a problem is to be detected by the unit of atomic locus. For example, employing *Postulate of Identity of Assigned Values-type2* (i.e., $\mathbf{P_{V2}}$) implying *continuity in attribute values*, the event *X* in (9-2) is to be inferred as (9-3) where the event causer z' can be different from z.

$$L(x,y,q_1,q_2,a,g,k)\bullet X\bullet L(z,y,q_3,q_4,a,g,k). \quad (9\text{-}2)$$

$$L(z',y,q_2,q_3,a,g,k). \quad (9\text{-}3)$$

## 9.2 Creation Problem Finding and Solving

Consider such a verbal command as S9-1 uttered by a human. Its interpretation is given by (9-4) as the goal event $X_G$ concerning the attribute, *Height* ($A_{03}$). If the current event $X_C$ is given by (9-5), then (9-6) with the transit event $X_T$ underlined can be inferred as the problem corresponding to S9-1.

(S9-1) Keep drone $C_8$ flying 7-9 meters high.

$$(\exists\ldots)L(z,C_8,q,q,A_{03},G_t,k)\wedge drone(C_8)\wedge 7m\le q\le 9m \quad (9\text{-}4)$$

$$(\exists\ldots)L(x,C_8,p,p,A_{03},G_t,k)\wedge drone(C_8) \quad (9\text{-}5)$$

$$(\exists\ldots)\underline{L(z_1,C_8,p,q,A_{03},G_t,k)}\bullet L(z,C_8,q,q,A_{03},G_t,k) \wedge drone(C_8) \wedge 7m\le q\le 9m \quad (9\text{-}6)$$

For this problem, the robot is to execute a task deploying a certain height sensor and actors, $z_1$ and $z$. The selection of the actor $z_1$ is performed as follows:

*If 9m-p <0 then* $z_1$ *is a sinker, otherwise*

*if 7m-p >0 then* $z_1$ *is a raiser, otherwise*

*7m≤p≤9m and no actor is deployed as* $z_1$.

The selection of *z* is a task in the case of the maintenance problem described below.

## 9.3 Maintenance Problem Finding and Solving

In general, the goal event $X_G$ for a maintenance problem is that for another creation problem, such as S9-1, possibly given by humans and solved by the robot in advance. That is, the task in this case is to autonomously restore the goal event $X_G$ created in advance to the current event $X_C$, as shown in (9-7), where the transit event $X_T$ is the reversal of such $X_{-T}$ that has been already identified as *abnormal* by the robot.

For example, if $X_G$ is given by (9-4) in advance, $X_T$ is also represented as the underlined part of (9-6) while $X_{-T}$ as (9-8). Therefore, the task here is quite similar to that which was described in the previous section.

$$X_G \bullet X_{-T} \bullet X_C \bullet X_T \bullet X_G \tag{9-7}$$

$$L(z_1, C_8, q, p, A_{03}, G_t, k) \wedge drone(C_8) \tag{9-8}$$

# 10
# Human Language Understanding by Robots

One day, Taro, smiling mischievously, says to Anna "You can see the package on the desk, can't you? That is a surprise present for Tom. Will you take it to him without notice?" Immediately, she replies "Ok! But, where does he live?" As he cannot remember Tom's exact address, he begins to locate Tom's home in words. After repeating question answering, she comes to a full understanding of the map that he describes.

## 10.1 Two-Staged Robotic Understanding of Human Language

In the situation above, Taro verbalizes his spatiotemporal mental image in natural language based on human-specific semantics and Anna tries to understand his utterances so as to actuate her sensors and motors appropriately for the purpose. The process of Anna's natural language understanding is two-staged. Once she interprets Taro's description into mental image description language (i.e., $\boldsymbol{L}_{md}$) expression based on human-specific semantics she then translates, it into $\boldsymbol{L}_{md}$ expression specific to her embodiment. This two-staged process enables her to think about people's utterances both in human-specific and robot-specific formal systems. That is, it is probable for her to be provided with artificial organs such as ultrasonic and infrared sensors that enable her to sense what people alone cannot, which may allow her to take paths different from people's in finding and solving problems. For example, consider such a case that Anna finds something pyramidal in a dark place through her ultrasonic sensor. She would never ask Taro to see the object but could say to him cautiously "Taro, there is something pyramidal in the dark." Her recognition of this situation is to be formulated as (10-1), reading that something $x$ is pyramidal at Anna's standard $K_A$ of shape ($A_{11}$) but of no value (/) at Taro's standard $K_T$. Thanks to her notice, he comes to share her perception of the object without seeing it for himself.

$$L(\_,x,Pyramidal,Pyramidal,A_{11},G_t,K_A)\Pi L(\_,x,/,/,A_{11},G_t,K_T). \quad (10\text{-}1)$$

It is noticeable that the formal system based on $\boldsymbol{L_{md}}$ permits coexistence of multiple values of an attribute with *different* standards (cf., *Postulates of Identity of Assigned Values* in Chapter 7).

## 10.2 Robotic Concept System for Intuitive Human-Robot Interaction

The author has a hypothesis that any distal stimulus to humans is first articulated into a set of gestalts as the sensory image in the brain subconsciously (i.e., intrinsic articulation), and then into a meaningful structure consciously (or purposively) by intuition and reason (i.e., semantic articulation). The distinction between intuition and reason is very important for robotic natural language understanding systems because most natural language expressions that people utter are intuitive and could embarrass robots. For example, consider a somewhat odd expression, such as S10-1, uttered by the people travelling in the train. It describes a certain intuitive impression of the scenery viewed from the running train. Its rational expression could be S10-2 but it would be understandable instantly by our daily experiences. How should robots process S10-1? Can robots reject it because it is unscientific? The most important fact is that ordinary people seldom find much serious inconvenience in their intuitive cognition of the world that leads to such a quite natural expression as S10-3. This refers to a psychological phenomenon of the same causality as S10-1, due to the observer's movement, but does not sound odd at all. On the contrary, S10-4 can be much more unnatural or unusual than S10-3 because people cannot observe the fact described by it. However, it is very usual for people to express their intuitive perception as directly as S10-1.

(S10-1) The station is running away from our train.

(S10-2) Our train is running away from the station.

(S10-3) The sun is moving from east to west of the earth.

(S10-4) The earth is rotating on its own axis from west to east around the sun.

This implies that a robot should be provided with a certain model of a human cognition system besides their own artificial cognition system in order to keep communication with people as comprehensible as possible. For this purpose, mental image directed semantic theory is to provide robots with each of these cognition systems as a formal system based on $\boldsymbol{L_{md}}$ and additionally with bi-directional translation rules between them. The formal system is not for assigning truth values to natural language expressions but interpretations as phenomena in sensory systems, namely, loci in attribute spaces. The observer-object relativity is denoted by the different coordinate frames with their origins at the parameter Standard. The scenes being described by S10-1 and S10-2 can be expressed as (10-2) and (10-3) in $\boldsymbol{L}_{md}$, respectively. Both the expressions are semantically plausible when the origin of the coordinate frame is assumed to be the location of *Train* for (10-2), namely, $k_{11}$ =*Train* and for (10-3), $k_{21}$≠*Train*. This is a matter of course for mental image directed semantic theory because all attribute values are relative to the standards.

$$(\exists\ldots)L(x_1,\text{Station},\text{Train},z_1,A_{12},G_t,k_{11})\Pi\ L(y_1,\text{We},\text{Train},\text{Train},A_{12},G_t,k_{12})\wedge z_1\neq\text{Train}. \quad (10\text{-}2)$$

$$(\exists\ldots)L(x_2,\text{Train},\text{Station},z_2,A_{12},G_t,k_{21})\Pi\ L(y_2,\text{We},\text{Train},\text{Train},A_{12},G_t,k_{22})\wedge z_2\neq\text{Station}. \quad (10\text{-}3)$$

Concerning intuitive expressions due to the observer-object relativity, we Japanese have a fantastic song for children, given below.

> "Come, the train's running, running, puffing, puffing, choo, choo, choo.
> With us on board, running, puffing, choo, choo, choo.
> So fast, so fast, outside the window, fields are ***flying away***, houses are ***flying away***.
> Run fast, run fast, run fast! The bridge's ***coming up***, ***coming up***. What a great fun!"
>
> ("Kisha-poppo", translated by Yokota)

## 10.3 Compound Concept System for Robots

Mental image directed semantic theory has presented a model of attention-guided perception yielding semantic articulation of the world to be modelled as loci in attribute spaces. Remarkably to mention, what and how are to be attended to is indicated in locus formulas because they are intrinsically the trajectories of the focus of attention of the observer. For example, the relativity of the observer's attention to its scanning object is recorded by Pattern and Standard parameters (i.e., $\omega_6$ and $\omega_7$ at (6-1)) and its scanning order for articulation is known by the temporal connectives involved.

For example, Anna is to interpret Taro's question, S10-5, as (10-4) in $\boldsymbol{L}_{md}$ according to human semantics.

(S10-5) What is between the table and the desk?

$$L(\_, \text{ISR}, \text{Table}, ?x, \text{A12},\text{Gs},\_)\bullet L(\_, \text{ISR}, ?x, \text{Desk}, \text{A12},\text{Gs},\_). \quad (10\text{-}4)$$

This $\boldsymbol{L}_{md}$ expression can suggest to Anna how to find the object in question (i.e., *?x*) so that it is located on the trajectory of the focus of Taro's attention running from the table to the desk. Then, she is to translate (10-4) systematically into another mental image description language expression with certain robotic semantics reflecting her embodiment that will enable her to work her sensory system so as to find *?x*.

Therefore, in order to remove such a kind of human-robot cognitive divide, robots have only to be provided with a certain capability to perform systematic bi-directional translation between $\boldsymbol{L}_{md}$ expressions yielded by human cognition and by robotic cognition. This idea inevitably requires us to clarify correspondence relations between natural concept system and artificial concept system shown in Fig. 10-1.

In this figure, the symbol $C_X$ represents the set of natural concepts of the world (W) for human (X=N) and the set of artificial concepts for robot (X=A). On the other hand, $S_X$ stands for the semantics of natural language as a subset of $C_X$. The two intelligent entities are deemed to form different concept systems due to their different embodiments, respectively. The cognitive robot must be able to perform a bidirectional translation between $S_N$ and $S_A$ in order to keep up good communication with people. That is, the cognitive robot can understand people's utterances both in the semantics $S_N$ and $S_A$.

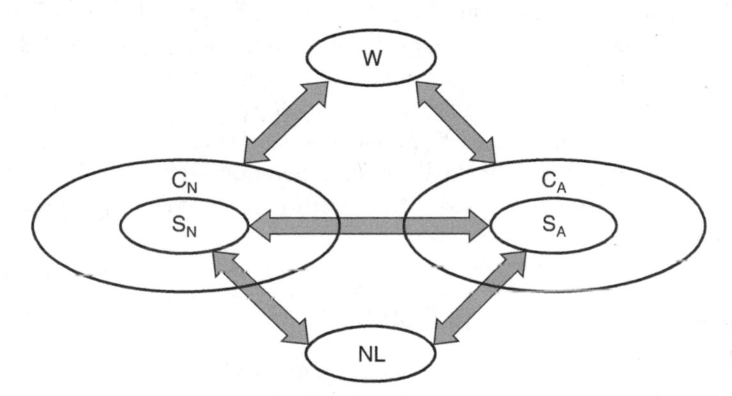


**Fig. 10-1.** Configuration of compound concept system for cognitive robots.
(W: world; NL: natural language; $C_N$ and $S_N$: natural concept system and semantics for human; $C_A$ and $S_A$: artificial concept system and semantics for robot; X⇔Y: correspondence between X and Y).

## 10.4 Robot Manipulation as Cross-Media Operation via Mental Image Description Language

Mental image directed semantic theory has been applied to several versions of the intelligent system IMAGES (e.g., Yokota, 2005), where robot manipulation is to be realized as cross-media operation via $\boldsymbol{L}_{md}$.

Among them, IMAGES-M is the prototype version of Conversation Management System (i.e., CMS) still under development and evolution for integrated multimedia understanding. This system, as shown in Fig. 10-2a, is one kind of expert system consisting of Inference Engine (IE), Knowledge Base (KB) and five kinds of user interface: (1) Text Processing Unit (TPU), (2) Speech Processing Unit (SPU),

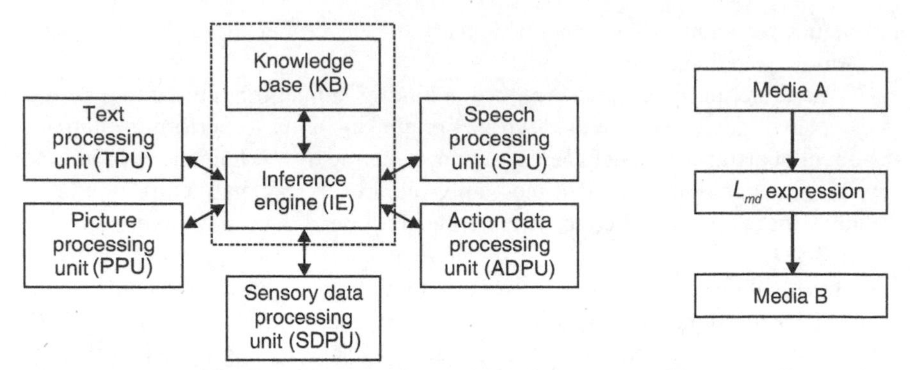


(a) Configuration of interlingual understanding model aiming at general purpose system-M.

(b) Cross-media operation.

**Fig. 10-2.** Interlingual understanding model aiming at general purpose system-M and cross-media operation via mental image description language ($\boldsymbol{L}_{md}$).

(3) Picture Processing Unit (PPU), (4) Action Data Processing Unit (ADPU) and (5) Sensory Data Processing Unit (SDPU). The pair of IE and KB work as Integration Processing Agent, Knowledge Processing Agent, and Emotion Processing Agent of the mind model shown in Fig. 4-1, and the group of user interfaces as Stimulus Processing Agent and Response Processing Agent. As depicted in Fig. 10-2b, these user interfaces are intended to convert information media and $\boldsymbol{L}_{md}$ expressions mutually in collaboration with IE and KB in order to facilitate various types of cross-media operations, such as language-to-picture translation.

## 10.5 Aware Computing in Robots

As shown in Table 5-2 and 5-3, ordinary people live their casual life attending to tens of attributes of the matters in the world in order to cognize them in comparison with several kinds of standards. They must, however, control their focus of attention quite efficiently according to their interest or so during their cognitive processes and this is also the case for the robot intended here, namely, the robot installed with IMAGES-M. Aware computing is conventionally conceived as sensory data computing driven by certain heuristics, cognitively motivated and intended to reduce their computational cost (Yokota, 2010). Awareness is tightly related to perception and attention (Searle, 1992; Koch, 2004). For example, Gestalt psychologists argue that people perceive gestalts in individual sensations, attend to them and become aware of what formed them. As for the robot, compared with the human mind, awareness of something can be defined as existence of any $\boldsymbol{L}_{md}$ expression about it on the Inference Engine. Nevertheless, it must be remarked that awareness in the robot is not always observable for us, only its report put out for us. For example, if the robot is aware of 'some big and red apple', we can know that indirectly by such a report as "There is a big red apple and this is its picture. Shall I bring it to you?".

In the natural course, aware computing in IMAGES-M is performed as computation on $L_{md}$ expressions in its Inference Engine. This computation is controllable by selecting the part or property of an $L_{md}$ expression to focus on, for example, Attribute, Standard, Pattern, etc. The most conventional method for awareness controlling in the robot is to employ its world knowledge. For example, if the robot is ordered to find a green apple, its knowledge of *green* and *apple* can focus its attention exclusively on the attributes *Color* and *Shape* of anything in its environment. Without such awareness controlling, robots would always have to pay full attention to everything in their environments. The semantic understanding of human verbal suggestion makes the robot abstractly (i.e., conceptually) aware of which matters and attributes involved in its sensations should be attended to, and its pragmatic understanding provides the robot with a concrete idea of real matters with real attribute values significant for its action. More exactly, semantic understanding in $L_{md}$ of human suggestion enables the robot to control its attention mechanism in such a top-down way that focuses the robot's attention on the significant attributes of the significant matters involved in its sensations, regardless of saliencies there. Successively, in order for pragmatic understanding in $L_{md}$ of human suggestion, the robot is to select the appropriate sensors corresponding with the suggested attributes and make them run on the suggested matters so as to pattern after the movements of human focus of attention implied by the locus formulas yielded in semantic understanding. That is, ***$L_{md}$ expression suggests to the robot what and how should be attended to in its environment.***

## 10.6 Homogeneous/Inhomogeneous Communication

In association with human and robotic concept systems, homogeneous communication is defined here as communication between cognitive entities equipped with sensory systems of the same functionality for knowledge acquisition from the world. Homogeneousness in communication is to be defined analytically based on mental image directed semantic theory. The sensory system of an entity can be characterized by the triplet <A,G,K>, denoting Attributes, Patterns, and Standards, respectively. As already mentioned, A, G, and K correspond with a set of sensors, spatiotemporal functionality, and value standards, respectively. If <A,G,K> is equal, then the interaction between the two entities X and Y is called homogeneous and otherwise inhomogeneous. In inhomogeneous cases, it is required that the information understanding facility of each with a translator between the pair of <A,G,K> different in each be provided. The difference between <A,G,K> is due to that between sensory systems. Attributes and values correspond with the sensors of the intelligent entity and their functionalities. That is, homogeneousness reflects the similarity between the bodily identities of the both entities, largely, species the entities belongs to, such as human, dog, and robot. The 4D world projected into the sensory system of the cognitive entity is to be described in $L_{md}$ with its specific <A,G,K>. For example, people can perceive neither ultraviolet as a value of the attribute 'color' nor ultrasound as a value of the attribute 'sound' but certain kinds of animals can. It is possible to hypothesize that the semantics of a sign system are characterized by its owner's sensory systems, more analytically, their competences and performances. Therefore, the families of attributes and the value ranges of an attribute can be different from species to species. When there is one-to-one correspondence between the two families of attributes, the communication between the two entities is called homogeneous. A cognitive entity can perceive the spatial and temporal extension of an attribute of a matter (i.e., object or event) in the 4D world. In general, the sensors such an entity possesses can work actively under the control by its attention mechanism triggered by external or internal stimuli over certain thresholds in strength. The spatial ranges its sensors can cover at a time are intrinsically limited and therefore it needs to make its sensors scan (or run about on) the objects which are too large for it to perceive at a glance. How should a sign system be identified in relation to its owner? In other words, natural language to human. It is quite difficult to count up all the values of an attribute that the entity can sense. Therefore, the values specific to the entity are to be denoted by the value of standard K specific to the entity. This scheme makes people permit such a conversation as follows. The participants C1 and C2 have different standards regarding the size of a book. Therefore, our natural language understanding system treats these utterances not to be contradictive each other but to be different in the standard values specific to the participants.

C1: " This book is large."

C2: " It is rather small."

Mental image directed semantic theory is in the same line with common coding theory, claiming that there is a shared representation (a common code) for both perception and action (Prinz, 1990, 1994) but it is quite remarkable that $L_{md}$ expression can be a common code as well for thought in the brain, namely, available for reasoning.

# 11

# 4D Language Understanding for Cognitive Robotics

As described in Chapter 8, the knowledge representation language (i.e., KRL), mental image description language (i.e., $L_{md}$), has a good capability to formulate human concepts expressed in 4D language as mental images. This chapter clarifies the requirements for robotic natural language understanding viewed from systematic computation and presents concrete algorithms to satisfy them, focusing on 4D language.

## 11.1 Semantic and Pragmatic Understanding of Natural Language

Spatiotemporal (i.e., 4D) language, as already mentioned, is the most important tool for intuitive human-robot interaction, as shown in Fig. 1-2. In such a situation, robots must understand natural language *semantically* and *pragmatically* well enough for comprehensible communication with humans. As already shown in Fig. 3-3, semantic understanding means associating symbols to conceptual images of matters, and pragmatic understanding means anchoring symbols to real matters by unifying conceptual images with perceptual images.

Robot manipulation by verbal suggestion is defined as human-robot interaction where a human gives a robot a verbal expression of his/her intention and the robot *behavioralizes* its conception, namely, the result of pragmatic understanding of the suggestion, as shown in Fig. 11-1. Here, behaviorize (or behaviorization) is the author's own special coinage whose meaning is to work sensors or effectors *actively*. As detailed above, semantic understanding is purely symbol manipulation for translation from verbal expression (text or speech) into $L_{md}$ expression and fundamental semantic computations on $L_{md}$ expressions by employing word meanings. On the other hand, pragmatic understanding is rather complicated because unstructured data processing for sensor-actor coordination is inevitably involved as well as pure symbol manipulation.

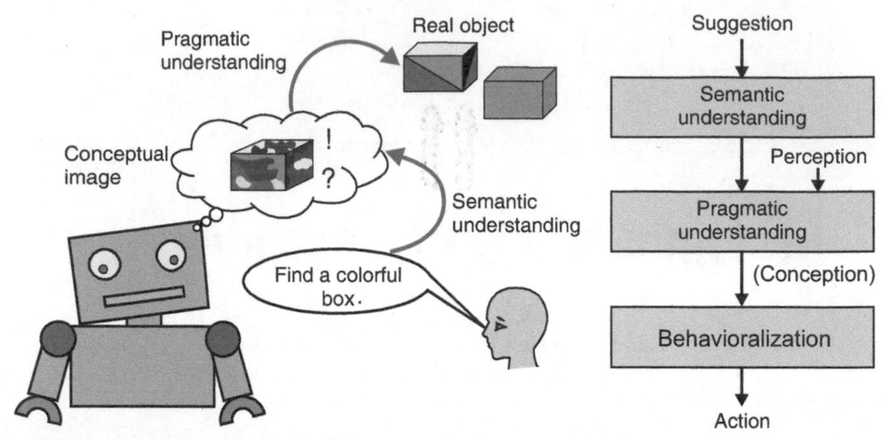


**Fig. 11-1.** Behavioralization of human suggestion by robots.

## 11.2 Requirements for Robotic Understanding of Natural Language

The author has already reported about integrative multimedia understanding based on $\boldsymbol{L}_{md}$ in the intelligent system IMAGES-M (e.g., Yokota, 2005), where its capability of natural language understanding is intended at least to satisfy the three requirements (R1-R3) for semantic understanding (Woods, 1975) and the two (R4 and R5) for pragmatic understanding as follows. It is notable that behavioralization is anchoring $\boldsymbol{L}_{md}$ on real behavior, namely, active operation or sensing of the physical environment. The last requirement R6 is seldom discussed but what the author thinks an ideal knowledge representation should be provided with.

(R1) Logical adequacy of $\boldsymbol{L}_{md}$

(R2) Translation between natural language and $\boldsymbol{L}_{md}$

(R3) Reasoning in $\boldsymbol{L}_{md}$

(R4) Anchoring via $\boldsymbol{L}_{md}$

(R5) Behavioralization via $\boldsymbol{L}_{md}$

(R6) Systematic interpretation of $\boldsymbol{L}_{md}$

The following sections detail how these requirements are satisfied in IMAGES-M.

## 11.3 Logical Adequacy of Mental Image Description Language

Ambiguity and vagueness are intrinsic to natural language. Therefore, it is essential for a KRL to provide a formal system with good capability, so called logical adequacy, in order to interpret any sentence into certain expressions in the KRL, uniquely explicating every possible case of ambiguity or vagueness.

As easily imagined, logical adequacy of $\boldsymbol{L}_{md}$ is mainly due to its expressive power, which in turn depends exclusively upon the set of constants specific to the mental image model because logical ones are granted for being rigid (i.e., without ambiguity or vagueness) and common to any logic-based KRL. In the formal system based on $\boldsymbol{L}_{md}$, every word concept is defined in the context of constants grounded in human sensation. In this sense, any word concept can be uniquely formulated but such word concepts for human sensation, such as *red* and *sweet*, forced to be constants instead of real sensation, are inevitably ambiguous or vague. The formal system is to explicate and inherit this kind of uncertainty in a context of Value, Attribute, and Standard, such as (sweet, Taste($A_{29}$)), Standard($k_1$)) and (sweet, Odor($A_{30}$), Standard($k_2$)). At present, about 50 kinds of Attribute concerning the physical world have been extracted exclusively from English and Japanese words as shown in Table 4-2. They are associated with all of the 5 senses (i.e., sight, hearing, smell, taste and touch) in our everyday life, with 7 kinds of Standard, as shown in Table 4-3.

Furthermore, mental image directed semantic theory has pointed out the necessity to discern spatial change events from temporal change events in robotic natural language understanding and has introduced the parameter Pattern (i.e., $g$) to $\boldsymbol{L}_{md}$, namely, $g=G_t$ for temporal change and $g=G_s$ for spatial change. As already mentioned above, most 4D function words, such as verbs and prepositions, can refer to both the event types and it is, therefore, crucial for robots to disambiguate natural language expressions containing these polysemous words, otherwise resulting in serious situations, as shown in Fig. 4-4. The intelligent sytem IMAGES conducts this kind of disambiguation by checking whether or not the word in dependency relation with such a 4D function word can be semantically unified. For example, consider S11-1 and 2 whose semantic contents are expressed as (11-1) with $g=G_s$ and with $g=G_t$, respectively. They are semantically ambiguous for robots because the concepts of the verb *run* and the preposition *via* are given as (11-2) and (11-3) with the two possibilities that $g=G_s$ and $g=G_t$. In each case, the success of semantic unification depends on whether or not the subject (*road* or *bus*) and the verb (*run*) or the preposition (*via*) have the common event pattern, namely, $L(\_,x,p,q,A_{12},G_s,\_)\wedge p\neq q$ for *road*($x$) and $L(\_,x,p,q,A_{12},G_t,\_)\wedge p\neq q$ for *bus*($x$). In other words, a road does lie over a long spatial range but does not displace itself in time, which is opposite to the case of a bus. The details of the semantic unification are given in Section 11.3 and 11.4 concerning the requirements R1 and R2.

(S11-1) The road runs from the city to the town via the village.

(S11-2) The bus runs from the city to the town via the village.

$(\exists\ldots)\underline{L(\_,x,x_1,x_2,A_{12},\{G_s/G_t\},\_)}\Pi((\underline{\underline{L(\_,x,x_3,y,A_{12},\{G_s/G_t\},\_)}}\bullet$

$$(\underline{L(\_,x,y,x_4,A_{12},\{G_s/G_t\},\_)})\wedge\{road(x)/bus\ (x)\}\wedge city(x_1)\wedge town(x_2)\wedge village(y). \quad (11\text{-}1)$$

$$(\lambda x)run(x)\Leftrightarrow(\lambda x)(\exists\ldots)L(\_,x,p,q,A_{12},g,\_)\wedge p\neq q\ (g=G_s \text{ or } G_t). \quad (11\text{-}2)$$

$$(\lambda x,y)via(x,y)\Leftrightarrow(\lambda x,y)(\exists\ldots)((L(\_,x,x_3,y,A_{12},g,\_)\bullet(L(\_,x,y,x_4,A_{12},g,\_)) \wedge x_3\neq y\wedge x_4\neq y. \quad (11\text{-}3)$$

The single-underlined part of (11-1) corresponds with '*x* runs from $x_1$ to $x_2$' and the double-underlined part, with '*x* via *y*'. This formula is to be reduced to (11-1') by application of the postulates and inference rules introduced in Chapter 7, where $x_3$ and $x_4$ are unified with $x_1$ and $x_2$, respectively.

$$(\exists\ldots)((L(\_,x,x_1,y,A_{12},\{G_s/G_t\},\_)\bullet L(\_,x,y,x_2,A_{12},\{G_s/G_t\},\_))\wedge\{road(x)/bus(x)\} \wedge city(x_1)\wedge town(x_2)\wedge village(y). \quad (11\text{-}1')$$

For another example, consider such somewhat complicated sentences as S11-3 and S11-4. The underlined parts are deemed to refer to some events neglected in time and in space, respectively. These events correspond to the skipping of the focus of attention of the observer and are called *Temporal Empty Event* and *Spatial Empty Event*, denoted by $\varepsilon_t$ and $\varepsilon_s$ as empty events with $g=G_t$ and $g=G_s$ at definition of empty event, respectively. The semantic interpretations of S11-3 and S11-4 are given by (11-4) and (11-5), where $A_{15}$ and $A_{17}$ represent the attribute *Trajectory* and *Mileage*, respectively.

(S11-3) The *bus* runs 50 km straight south from A to B, and _after a while_, at C it meets the street with the sidewalk.

(S11-4) The *road* runs 50 km straight south from A to B, and _after a while_, at C it meets the street with the sidewalk.

$$(\exists\ldots)(L(\_,x,A,B,A_{12},G_t,\_)\Pi L(\_,x,0,50\text{ km},A_{17},G_t,\_)\Pi L(\_,x,Point,Line,A_{15},G_t,\_)\Pi L(\_,x,South,South,A_{13},G_t,\_))\bullet\varepsilon_t\bullet(L(\_,x,p,C,A_{12},G_t,\_)\Pi L(\_,y,q,C,A_{12},G_s,\_)\Pi L(\_,z,y,y,A_{12},G_s,\_))\wedge bus(x)\wedge street(y)\wedge sidewalk(z)\wedge p\neq q. \quad (11\text{-}4)$$

$$(\exists\ldots)(L(\_,x,A,B,A_{12},G_s,\_)\Pi L(\_,x,0,50\text{ km},A_{17},G_s,\_)\Pi L(\_,x,Point,Line,A_{15},G_s,\_)\Pi L(\_,x,South,South,A_{13},G_s,\_))\bullet\varepsilon_s\bullet(L(\_,x,p,C,A_{12},G_s,\_)\Pi L(\_,y,q,C,A_{12},G_s,\_)\Pi L(\_,z,y,y,A_{12},G_s,\_))\wedge road(x)\wedge street(y)\wedge sidewalk(z)\wedge p\neq q. \quad (11\text{-}5)$$

From the viewpoint of cross-media reference as integrative multimedia understanding, the formula (11-5) can refer to such a spatial change event depicted as a still picture while (11-4) can be interpreted into a motion picture (Yokota and Capi, 2005a).

A considerable number of spatial terms have been analyzed over various kinds of English words, such as prepositions, verbs, adverbs, etc., categorized as *Dimensions*, *Form* and *Motion* in the class *SPACE* of the Roget's thesaurus (Roget,

1975), and found that almost all the concepts of spatial change events can be defined in exclusive use of five kinds of attributes for the focus of attention of the observer, namely, *Physical location* ($A_{12}$), '*Direction* ($A_{13}$)', *Trajectory* ($A_{15}$), *Mileage* ($A_{17}$) and *Topology* ($A_{44}$). This fact as well is evidence proving that the natural semantics (i.e., semantics of natural language) are quite subjective to people's cognitive processes.

## 11.4 Translation Between Natural Language and Mental Image Description Language

Natural language and $\boldsymbol{L}_{md}$ are to be translated mutually in use of word meaning definition. A word meaning $M_W$ is defined as a pair of *Concept Part* ($C_p$) and *Unification Part* ($U_p$) and is formulated as (11-6).

$$M_W = [C_p{:}U_p]. \tag{11-6}$$

The $C_p$ of a word $W$ is an $\boldsymbol{L}_{md}$ expression as its concept while its $U_p$ is a set of operations for unifying the $C_p$s of $W$'s syntactic governors or dependents. There are three types of commands employed for unification operations: ARG($Z_1$, $Z_2$), PAT($Z_1$,$Z_2$,$Z_3$) and LOG($Z_1$,$Z_2$,$Z_3$). These are controlled in application by the condition command COND($Z_1$,$Z_2$). Their details are as follows.

i) ARG($Z_1$, $Z_2$)
This command indicates the system to substitute the argument of $Z_1$ to $Z_2$ for their unification.

ii) PAT($Z_1$,$Z_2$,$Z_3$)
Formulas $Z_1$ and $Z_3$ are combined with a sub-formula $Z_2$ included in the both common, where $Z_1$, $Z_2$, and $Z_3$ are all locus formulas. $Z_2$ is called 'Unification Handle ($U_h$)'. For example, given $Z_1$=A$\Pi$B, $Z_3$=B•C and $Z_2$=B, then PAT($Z_1$,$Z_2$,$Z_3$)=(A$\Pi$B)•C. For another example, given $Z_1$=C•B, $Z_3$=B$\Pi$A and $Z_2$=B, then PAT($Z_1$,$Z_2$,$Z_3$)=C•(B$\Pi$A). One of $Z_1$ and $Z_3$ is the concept part of the current word, denoted by $, and if the other is that of the main governor in the clause, it is denoted by #.

iii) LOG($Z_1$,$Z_2$,$Z_3$)
This indicates the same operation as PAT, except that the arguments are not locus formulas but constituents combined with the purely logical connectives, such as AND, OR, etc., as in $C_1(x) \wedge C_2(y)$. For example, given $Z_1$=A$\wedge$B, $Z_3$=B$\wedge$C, and $Z_2$=B, then LOG($Z_1$,$Z_2$,$Z_3$)=A$\wedge$B$\wedge$C where all the arguments are assumed as conjunctive normal forms.

iv) COND($Z_1$,$Z_2$)
This denotes such a condition that the part of speech of $Z_1$ is $Z_2$. The three operations ARG, PAT and LOG are to appear in the places of $O_i$'s such a context that COND($Z_1$,$Z_2$)$\rightarrow O_1, O_2, \ldots, O_n$. Each operation command in this form is executed by the system if and only if the condition is absent or true. For example, the meaning of the English verb *carry* can be given by (11-7).

$[(\lambda x,y)(\exists p,q,k)L(x,\{x,y\},p,q,A_{12},G_t,k)\wedge x\neq y\wedge p\neq q$: ARG(Dep.1,x); ARG(Dep.2,y); LOG($,∧,Dep.1); LOG($,∧,Dep.2);]. (11-7)

The $U_p$ above consists of two operations to unify the first dependent (*Dep.1*) and the second dependent (*Dep.2*) of the current word with the variables *x* and *y*, respectively. Here, *Dep.1* and *Dep.2* are the 'subject' and the 'object' of 'carry', respectively. Therefore, the sentence '*Mary carries a book*' is translated into (11-8).

$$(\exists\ldots)\ L(Mary,\{Mary,y\},p,q,A_{12},G_t,k)\wedge Mary\neq y\ \wedge p\neq q\ \wedge book(y). \quad (11\text{-}8)$$

Figure 11-2 shows the details of the conversion process of a surface structure (text) into a conceptual structure (text meaning) through a surface dependency structure (or tree).

For another example, the meaning description of the English preposition 'through' is also given by (11-9).

$[(\lambda x,y)(\exists\ldots)\ (\underline{L(x,y,p_1,z,A_{12},g,k)}\bullet L(x,y,z,p_3,A_{12},g,k))\Pi\ L(x,y,p_4,p_4,A_{13},g,k_0)$
$\wedge p_1\neq z\wedge z\neq p_3$:ARG(Dep.1,z);LOG(#,∧,Dep.1);
COND(Gov,Verb)→ PAT(Gov,(1,1),$); (11-9)
COND(Gov,Noun)→ARG(Gov,y), PAT(#,Π,$);] .

The $U_p$ above is for unifying the $C_p$s of the very word, its governor (*Gov*, a verb or a noun) and its dependent (*Dep.1*, a noun). The second argument (1,1) of the command *PAT* (i.e., $U_h$) indicates the underlined part of (11-9) and, in general, (*i,j*) refers to the partial formula covering from the *i*th to the *j*th atomic formula of the current $C_p$. When $U_h$ is missing, the $C_p$s are to be combined simply with ∧. In the case that the unification indicated in a $U_p$ fails, whether or not for the natural language understanding system to reject the input natural language expression depends on its design. For example, in our natural language understanding system,

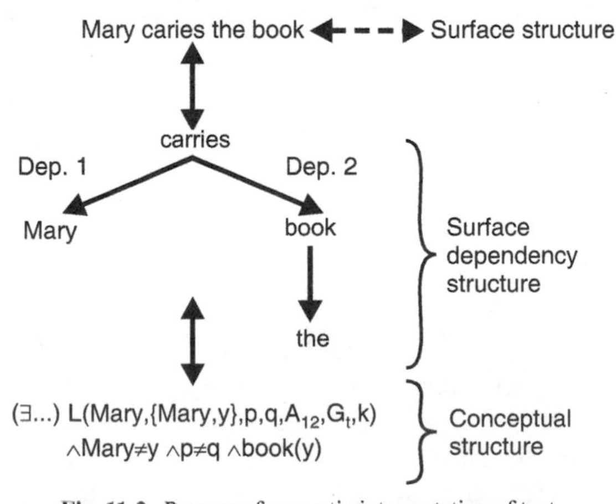


**Fig. 11-2.** Process of semantic interpretation of text.

the $U_p$ of 'through' is to fail for such expressions as 'Tom *stays* through the forest' and 'Mary *stops* through the tunnel'. If it is designed so as to be robust, it can try to reach some plausible interpretation of the input by employing deductive or non-deductive inferencing facilities.

Therefore, the sentences S11-5–S11-7 are interpreted as (11-10)–(11-12), respectively. The underlined parts of these formulas are the results of *PAT* operations. The expression (11-13) is the $C_p$ of the adjective *long*, implying that *there is some value greater than some standard of Length ($A_{02}$)*, which is often simplified as (11-14).

(S11-5) The train runs through the tunnel.

$$(\exists\ldots)\ (\underline{L(x,y,p_1,z,A_{12},G_t,k)\bullet L(x,y,z,p_3,A_{12},G_t,k)})$$
$$\Pi\ L(x,y,p_4,p_4,A_{13},G_t,k_0)\ \wedge p_1\neq z\ \wedge z\neq p_3\ \wedge train(y)\ \wedge tunnel(z). \quad (11\text{-}10)$$

(S11-6) The path runs through the forest.

$$(\exists\ldots)\ (\underline{L(x,y,p_1,z,A_{12},G_s,k)\bullet L(x,y,z,p_3,A_{12},G_s,k)})\Pi$$
$$L(x,y,p_4,p_4,A_{13},G_s,k_0)\ \wedge p_1\neq z\ \wedge z\neq p_3\ \wedge path(y)\ \wedge forest(z). \quad (11\text{-}11)$$

(S11-7) The path through the forest is long.

$$(\exists\ldots)\ (L(x,y,p_1,z,A_{12},G_s,k)\bullet L(x,y,z,p_3,A_{12},G_s,k))\Pi$$
$$L(x,y,p_4,p_4,A_{13},G_s,k_0)\Pi L(x_1,y,q,q,A_{02},G_t,k_1)\wedge p_1\neq z\ \wedge z\neq p_3\wedge q>k_1\wedge path(y)\wedge forest(z). \quad (11\text{-}12)$$

$$(\exists\ldots)L(x_1,y_1,q,q,A_{02},G_t,k_1)\wedge q>k_1. \quad (11\text{-}13)$$

$$(\exists\ldots)L(x_1,y_1,Long,Long,A_{02},G_t,k_1). \quad (11\text{-}14)$$

The CONDs in a Unification Part are designed to explicate ambiguity due to surface dependency structure. For example, S11-8 alone has two plausible interpretations (11-15) and (11-16) different at the underlined parts, implying 'Jack with the stick' and 'Tom with the stick', respectively. The meaning of *with* is defined as (11-17), where *Gdep.1* denotes *Dep.1* of *Gov*, that is, *Tom*, the subject of *follows* here.

(S11-8) Tom follows Jack with the stick.

$$(\exists\ldots)(L(Jack,Jack,p,q,A_{12},G_t,\_)\underline{\Pi L(Jack,x,Jack,Jack,A_{12},G_t,\_)})\bullet L(Tom,Tom,p,q,$$
$$A_{12},G_t,\_)\wedge p\neq q\ \wedge stick(x). \quad (11\text{-}15)$$

$$(\exists\ldots)L(Jack,Jack,p,q,A_{12},G_t,\_)\bullet(L(Tom,Tom,p,q,A_{12},G_t,\_)$$
$$\underline{\Pi L(Tom,x,Tom,Tom,A12,Gt,\_)})\wedge p\neq q\ \wedge stick(x). \quad (11\text{-}16)$$

$$[(\lambda x,y)(\exists\ldots)\ L(x,y,x,x,A_{12},g,k): ARG(Dep.1,y);$$

$$COND(Gov,Verb)\rightarrow ARG(Gdep.1,x);\ COND(Gov,Noun)\rightarrow ARG(Gov,x);]. \quad (11\text{-}17)$$

In general, dependency relations are classified into two groups: Obligatory and optional dependencies. The dependency between a verb and its subject (e.g.,

Tom←goes) or between a preposition and its object (e.g., to→London) belong to the former group and that between an noun and its adjective modifier (e.g., big←box) or between a verb and its preposition (phrase) modifier (e.g., travel →through(→America)), to the latter. According to the author's investigation, obligatory and optional dependencies can be made to correspond systematically with $\boldsymbol{L}_{md}$ expressions by the unification parts of the governors and the dependents, respectively. For example, the unification part of 'through' in (11-9) is defined so as to allow 'through U.S.A.' to be unified with 'travel' but not with 'stay'.

## 11.5 Reasoning in Mental Image Description Language

The fundamental semantic computations on $\boldsymbol{L}_{md}$ expressions are performed to detect semantic anomalies, ambiguities and paraphrase relations in texts. Semantic anomaly detection is very important for eliminating meaningless computations. For example, consider such a conceptual structure as (11-18), where $A_{39}$ is the attribute *Vitality*. This locus formula can correspond to the English expression 'The chair is alive', which is usually semantically anomalous because a *chair* never has vitality in the real world projected into the attribute spaces.

$$(\exists x)L(\_,x,Alive,Alive,A_{39},G_t,\_)\wedge chair(x). \qquad (11\text{-}18)$$

This kind of semantic anomaly can be detected in the following process. Firstly, assume the concept of *chair* as (11-19), where $A_{29}$ refers to the attribute *Taste*, reading that a chair never has either *Vitality* or *Taste*, and that...

$$(\lambda x)\ chair(x) \Leftrightarrow (\lambda x)\ (\ldots L^*(\_,x,/,/,A_{29},G_t,\_) \wedge \ldots \wedge L^*(\_,x,/,/,A_{39},G_t,\_) \wedge \ldots). \qquad (11\text{-}19)$$

Secondly, the trivial theorem (11-20) and the postulate *Postulate of Identity of Assigned Values* are utilized. The formula (11-20) means that ***if one of two loci exists in every time interval, then they can coexist***. Furthermore, postulate of identity of assigned values-type1 states that ***a matter never has different values of an attribute with a standard at a time***.

$$\chi_1 \wedge \chi_2{}^* .\supset. \chi_1\ \Pi\ \chi_2. \qquad (11\text{-}20)$$

Lastly, the semantic anomaly of 'live chair' is detected by using (11-18)–(11-20) and postulate of identity of assigned values-type1. That is, the formula (11-21) below is finally deduced from (11-18) and violates the commonsense given by postulate of identity of assigned values-type1, that is, *Alive≠/*.

$$(\exists x)L(\_,x,Alive,Alive,A_{39},G_t,\_)\ \Pi\ L(\_,x,/,/,A_{39},G_t,\_). \qquad (11\text{-}21)$$

This process above is also employed for dissolving such a syntactic ambiguity as found in S11-9. That is, the semantic anomaly of 'alive desk' is detected and eventually 'alive insect' is adopted as a plausible interpretation.

(S11-9) Look at the insect on the desk, which is still alive.

If a text has multiple plausible interpretations, it is semantically ambiguous. For example, S11-10 alone has two plausible interpretations, (11-22) and (11-23), different at the underlined parts, implying 'Jack with the stick' and 'Tom with the stick', respectively.

(S11-10) Tom follows Jack with the stick.

$$(\exists x)(L(Jack,Jack,p,q,A_{12},G_t,\_)\underline{\Pi L(Jack,x,Jack,Jack,A_{12},G_t,\_)})\bullet L(Tom,Tom,p,q,A_{12},G_t,\_)\wedge p\neq q. \quad (11\text{-}22)$$

$$(\exists x)L(Jack,Jack,p,q,A_{12},G_t,\_)\bullet(L(Tom,Tom,p,q,A_{12},G_t,\_)\underline{\Pi L(Tom,x,Tom,Tom,A_{12},G_t,\_)})\wedge p\neq q \wedge stick(x). \quad (11\text{-}23)$$

Among the fundamental semantic computations, detection of paraphrase relations is the most essential because it is for detecting equalities in semantic descriptions, the other two are for detecting inequalities in them. In the deductive system intended here, if two different texts are interpreted into the same locus formula, they are paraphrases of each other. For example, the sentence '*Mary goes with a book*' is interpreted into (11-24) which is proved to be equivalent to (11-8), the semantic description of '*Mary carries a book*'. In the process of this proof, (11-25) and (11-26) are utilized which are deduced from the syntax rules of $\boldsymbol{L}_{md}$.

$$(\exists\ldots)L(Mary,Mary,p,q,A_{12},G_t,k)\Pi L(Mary,y,Mary,Mary,A_{12},G_t,k)\wedge p\neq q\wedge book(y). \quad (11\text{-}24)$$

$$(\forall\ldots)\ (L(x_1,x_2,p,q,a,g,k)\Pi\ L(x_3,x_4,x_2,x_2,a,g,k).\supset_0.L(x_1,x_2,p,q,a,g,k)\Pi L(x_3,x_4,p,q,a,g,k)). \quad (11\text{-}25)$$

$$(\forall\ldots)(L(x_1,x_2,p,q,a,g,k)\Pi L(x_1,x_3,p,q,a,g,k)\ .\equiv_0.L(x_1,\{x_2,x_3\},p,q,a,g,k)). \quad (11\text{-}26)$$

For another example, S11-11 and S11-12 below can be proved to be paraphrases of each other by employing ***Postulate of Reversibility of Spatial Event***. That is, according to the reversal operations defined by definition of reversal operation of spatial change event ($\mathbf{D}_{\mathbf{RO}}$) or (8-34) introduced in Chapter 8, (11-27) is transformed into (10-28) as its reversal and is equivalent to the semantic interpretation of S11-12. These two formulas can be compared to the images recorded by the camera work shown in Fig. 11-3a and b, respectively. Remember Fig. 1-4 which implies that perception or recognition can be different due to different movements of the focus of attention of the observer upon the same scene.

(S11-11) The road separates at C from the street with the sidewalk, and after a while, runs 10 km straight west from B to *A*.

(S11-12) The road runs 10 km straight east from A to B, and after a while, at C it meets the street with the sidewalk.

$$(\exists\ldots)(L(\_,x,C,p,A_{12},G_s,\_)\Pi L(\_,y,C,q,A_{12},G_s,\_)\Pi L(\_,z,y,y,A_{12},G_s,\_))\bullet\varepsilon_s\bullet$$
$$(L(\_,x,B,A,A_{12},G_s,\_)\Pi L(\_,x,0,10\ km,A_{17},G_s,\_)\Pi L(\_,x,Point,Line,A_{15},G_s,\_)$$

$\Pi$L(_,x,West,West,$A_{13}$,$G_s$,_)) $\wedge$road(x)$\wedge$street(y)$\wedge$sidewalk(z)$\wedge$p$\neq$q. (11-27)

($\exists$...)(L(_,x,A,B,$A_{12}$,$G_s$,_)$\Pi$L(_,x,0,10 km,$A_{17}$,$G_s$,_)$\Pi$ L(_,x,Point,Line,$A_{15}$,$G_s$,_)$\Pi$ L(_,x,East,East,$A_{13}$,$G_s$,_))$\bullet\varepsilon_s\bullet$(L(_,x,p,C,$A_{12}$,$G_s$,_)$\Pi$L(_,y,q,C,$A_{12}$,$G_s$,_)$\Pi$ L(_,z,y,y,$A_{12}$,$G_s$,_))

$\wedge$road(x)$\wedge$street(y)$\wedge$sidewalk(z)$\wedge$p$\neq$q. (11-28)

## 11.6 Anchoring via Mental Image Description Language

As also mentioned above, an event expressed in $L_{md}$ is compared to a movie recorded through a floating camera because it is necessarily grounded in the movement of the focus of attention of the observer over the event. ***This implies that*** $L_{md}$ ***expression can suggest to the robot what and how should be attended to in its environment.*** For example, consider such a suggestion as S11-13 presented to the robot by a person. In this case, unless the robot is aware of the existence of a certain box between the stool and the desk, such semantic understanding of the underlined part as (11-29) and such a semantic definition of the word *box* as

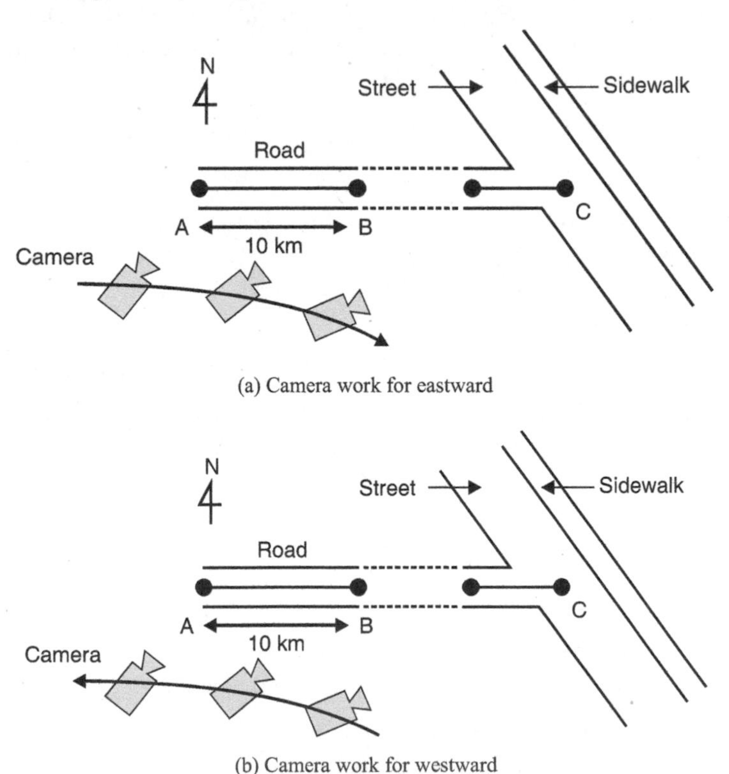


(a) Camera work for eastward

(b) Camera work for westward

**Fig. 11-3.** Images of spatial change events and camera works.

(11-30) are very helpful for it. The attributes $A_{12}$ (*Location*), $A_{13}$ (*Direction*), $A_{32}$ (*Color*), $A_{11}$ (*Shape*) and the spatial change event on $A_{12}$ in these $\boldsymbol{L}_{md}$ expressions indicate that ***the robot has only to activate its vision system in order to search for the box from the stool to the desk*** during the pragmatic understanding. That is, the robot can control its attention mechanism in the top-down way indicated in $\boldsymbol{L}_{md}$ to deploy its sensors or actuators during pragmatic understanding.

(S11-13) Avoid the green box between the stool and the desk.

$$(\exists \ldots)(\underline{L(\_,x_4,x_1,x_2,A_{12},G_s,\_)\bullet L((\_,x_4,x_2,x_3,A_{12},G_s,\_))\Pi L(\_,x_4,p,p,A_{13},G_s,\_)}\Pi$$
$$L(\_,x_2,Green,Green,A_{32},G_t,\_)\wedge stool(x_1)\wedge box(x_2)\wedge desk(x_3)\wedge ISR(x_4). \quad (11\text{-}29)$$

$$(\lambda x)box(x) \Leftrightarrow (\lambda x)L(\_,x,Hexahedron,Hexahedron,A_{11},G_t,\_)\wedge container(x). \quad (11\text{-}30)$$

The attributes listed in Table 4-2 are essentially for human sensors and, therefore, the locus formula as the result of semantic understanding should be translated into its equivalent concerning the attributes specific to the robot's sensors. For example, the value *Green* of *Color* ($A_{32}$) specified by the human should be paraphrased into its corresponding value at a certain light sensor specific to the robot, that is, subject to the robot. Such a knowledge piece is called Attribute Paraphrasing Rule (Yokota and Capi, 2005a, 2005b; Yokota, 2012).

## 11.7 Behavioralization via Mental Image Description Language

The process for behavioralization is to translate a conception into an action as an appropriate sequence of control codes for certain sensors or actuators in the robot to be decoded into a real behavior. For this purpose, there are two kinds of core procedures required, so called *locus formula paraphrasing* and *behavior chain alignment*.

In the same way as anchoring, the locus formula as the conception should be translated into its equivalent concerning the attributes specific (i.e., subjective) to the robot's sensory system. For example, an atomic locus of the robot's *Shape* ($A_{11}$) specified by the human should be paraphrased into a set of atomic loci of the *Angularity* ($A_{45}$) of each joint in the robot. For another example, *Velocity* ($A_{16}$) for the human into a set of change rates in *Angularity* ($A_{45}$) over *Duration* ($A_{35}$) (i.e., $A_{45}/A_{35}$) of the robot's joints involved. These knowledge pieces also belong to the set of attribute paraphrasing rules.

Ideally, the atomic loci in the conception (original or paraphrased) should be realized as the action in a perfect correspondence with an appropriate chain of sensor or actuator deployments. However, such a chain as a direct translation of the conception must often be aligned to be feasible for the robot due to the situational, structural or functional differences between the human and the robot. For an example of situational difference, consider the suggestion S11-14 to the robot Robby from the human Tom in the scenario in Fig. 11-4. In this case, the robot must interpolate the travel from its initial location to the green box and the action to pick up the box. This performance is in accordance with the principle of problem finding and solving described in Chapter 9.

(S11-14) Robby, carry the green box to the table.

On the other hand, for example of structural or functional difference, consider the case of behavioralization by a non-humanoid robot. Consider a verbal command, such as S11-15, uttered to the SONY AIBO robot, named *Poci*.

(S11-15) Poci, walk forward and wave your left hand.

The robot pragmatically understood the suggestion as '*I* walk forward and wave *my* left *foreleg*' based on the knowledge piece that only forelegs can be waved' and behavioralized its conception as 'I walk forward *BEFORE* sitting down *BEFORE* waving my left foreleg' but not as 'I walk, *SIMULTANEOUSLY* waving my left foreleg', in order not to fall down. The procedure involved for this alignment, based on the principle of problem finding and solving introduced in Chapter 9, is as follows.

Firstly, late in the process of cross-media translation from text to AIBO's action, this command is to be interpreted into (11-31) with the attribute S*hape* ($A_{11}$) and its values, *Walkf-1* and so on, at the standard of *AIBO*, reading that Poci makes himself walk forward and waves his left hand. Each action in an AIBO is defined as an ordered set of shapes (i.e., time-sequenced snapshots of the action) corresponding uniquely with the positions of their actuators, determined by the rotations of the joints. For example, the actions 'walking forward (***Walkf***)' and 'waving left foreleg (***Wavelf***)' are defined as (11-32) and (11-33), respectively.

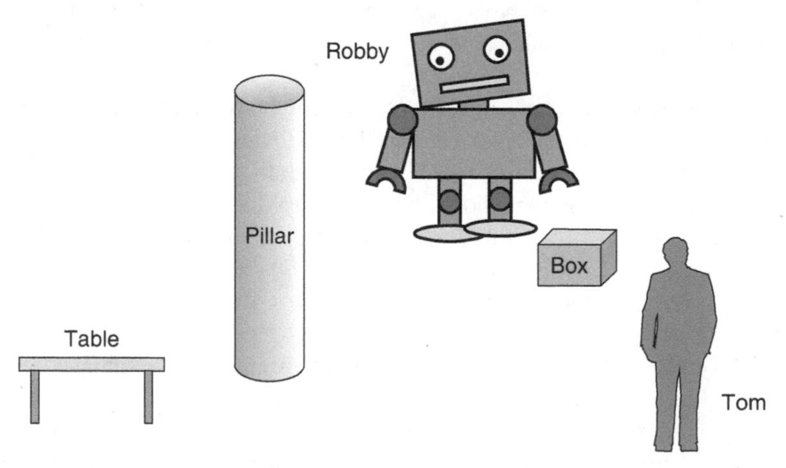


**Fig. 11-4.** A Scene of Tom and Robby.

**Fig. 11-5.** Poci's reaction to the command "Poci, walk forward and wave your left hand."

$$L(Poci,Poci,Walkf\text{-}1,Walkf\text{-}m,A_{11},G_t,AIBO)\wedge L(Poci,Poci,Wavelf\text{-}1,Wavelf\text{-}n,A_{11},G_t,AIBO). \quad (11\text{-}31)$$

$$\mathbf{W}alkf=\{Walkf\text{-}1,Walkf\text{-}2,\ldots,Walkf\text{-}m\} \quad (11\text{-}32)$$

$$\mathbf{W}avelf=\{Wavelf\text{-}1,Wavelf\text{-}2,\ldots,Wavelf\text{-}n\} \quad (11\text{-}33)$$

Secondly, an AIBO cannot perform the two events (i.e., actions) simultaneously and, therefore, the transit event between them is to be inferred as the underlined part of (11-34), the goal event ($X_G$) here.

$$L(Poci,Poci,Walkf\text{-}1,Walkf\text{-}m,A_{11},G_t,AIBO)\bullet \underline{L(Poci,Poci,Walkf\text{-}m,Wavelh\text{-}1,A_{11},G_t,AIBO)}\bullet L(Poci,Poci,Wavelf\text{-}1,Wavelf\text{-}n,A_{11},G_t,AIBO) \quad (11\text{-}34)$$

Thirdly, (11-35) is to be inferred, where the transit event, underlined, is interpolated between the current event $X_C$ (=(11-36)) and the goal event $X_G$ (=(11-34)).

$$X_C\bullet \underline{L(Poci,Poci,p2,Walkf\text{-}1,A_{11},G_t,AIBO)}\bullet X_G \quad (11\text{-}35)$$

$$L(Poci,Poci,p1,p2,A_{11},G_t,AIBO) \quad (11\text{-}36)$$

Finally, (11-35) is translated into a series of the joint angles in Poci, namely, the $\boldsymbol{L}_{md}$ expression with the semantics specific to AIBO. Figure 11-5 shows the reaction to the command performed by Poci.

## 11.8 Systematic Interpretation of Mental Image Description Language

Conventionally, KRLs for natural language understanding, including any semantic representation schemes, are seldom explicitly given the semantics of themselves but often implicitly, by employing natural language words in the vocabularies. That is, they owe their semantics to the conceptual images which the human readers have evoked by their lexical knowledge. On the contrary, $\boldsymbol{L}_{md}$ is explicitly provided with the semantics and systematically grounded in human mental images modeled as loci in attribute spaces. An attribute space is the model of a human sensory field, each of whose coordinates is assumed to correspond to a sensation. That is, the semantics of $\boldsymbol{L}_{md}$ are basically free of any specific natural language.

An assignment of meaning to the symbols of a formal language is called an *interpretation*. In general, formal languages are defined in solely syntactic terms, especially, in the fields of mathematics, logic, and theoretical computer science, and therefore, as already pointed out in 1.2, they do not have any meaning until they are given some *interpretation*. In any formal language, logical constants are always given the same meaning by every interpretation of the standard kind, so that only the meanings of the non-logical symbols are changed. Logical constants, as listed in 6.2, include quantifier symbols '∀'and '∃', symbols for logical connectives '∧', '∨', '~', and auxiliary constants '.', '(', ')'.

An interpretation of a formal language is conventionally based on the set theory (Frege, 1879) and given over a non-empty set *U* of individual constants called a universe (or domain) of discourse. An example of formal interpretation of the first-order language is as follows, where the description following '=' in [=...] is the intuitive interpretation of the symbol in natural language.

- Universe of discourse: U={a,b,c,d} [=Domestic animal world]
- Individual constants: a [=The white cat], b [=The iridescent rooster], c [=The black dog], d [=The white mouse]
- Predicate constants: B, C, I, J, K used in such forms as follows.

  B(x) [=*x* is a bird]
  C(x) [=*x* is a mammal]
  I(x) [=*x* is black]
  J(x) [=*x* is white]
  K(x, y) [=*x* is more colorful than *y*]

If a given interpretation assigns the value *True* to a sentence (or assertion) or a set of sentences (called a theory), the interpretation is called a *model* of that sentence or theory. In the interpretation above, *B*(*b*), *C*(*a*), *C*(*c*), *C*(*d*), *I*(*c*), *J*(*a*), *J*(*d*), *K*(*b*,*a*), *K*(*b*,*c*) and *K*(*b*,*d*) are true sentences, and the others, such as *B*(*a*), are false sentences. However, it is noticeable that without the intuitive interpretation in natural language, the interpretation would be nonsensical to people. The intuitive interpretation is more conventionally given in such a direct way as *Bird*(*b*), *Mammal*(*a*), and *Black*(*c*) on the assumption that human readers can understand this kind of notation in natural language of meta-use, that is, without any further explanation of the involved natural language words. Such a scheme may be viable in a very small domain with a very limited vocabulary and very trivial semantic relations among them, such as '$(\forall x)Mouse(x) \supset Mammal(x)$' and '$(\forall x)Water(x). \supset .Liquid(x) \wedge Clear(x) \wedge Tasteless(x)$'. This type of logical expression as is can give only combinations of dummy tokens at best. For example, it is no problem for machines to replace *bring*(*x*,*y*) and *come*(*y*) with *w013*(*x*,*y*) and *w025*(*y*) in such an assertion as '$(\forall x,y)bring(x,y) \supset come(y)$'. The latter do not represent any word concepts or meanings at all but are the *coded names* for the former. The very inconvenience you find in this kind of substitution is due to being without symbol grounding (Harnad, 1990) on your lexical knowledge of English. However, machines can pretend as if they understood natural language, if provided with bidirectional translation rules between natural language and the KRL such as '*x* bring *y* ←→ *w013(x,y)*' and '*y* come ←→ *w025(y)*', especially, which is the very case for the machines whose inputs and responses are limited to natural language expressions and which can answer a question such as 'When Tom brought the wine, did it come?' correctly. This approach to natural language understanding is called naïve semantics and remains without grounding in the non-linguistic real world. A naïve semantic theory is based upon a particular language, its syntax and its word senses, involving the use of a lexical theory, which maps each word sense to a simple theory (or set of assertions) about the objects or events of reference.

There are no issues in employing naïve semantics without considering each word meaning when the requirements for a semantic theory issued by Katz and Fodor (Katz and Fodor, 1963) are neglected. For example, if A(x) and B(x) contradict each other, it is possible to introduce such an axiom as ~(A(x)∧B(x)), good for ~(Dog(x)∧Cat(x)) but not for ~(Dog(x)∧Mammal(x)), resulting in a tremendously numerous set of axioms concerning the semantic relations among words. That is, these simple-minded translations are possible only within a carefully selected, limited vocabulary where no semantic problems occur.

As for another problem, any AI approach to semantics has the risk of circular definition of word meanings, especially in a large domain. For example, the following descriptions are some examples of circular definitions found in Webster's, where the first entry words are the verbs 'drive' and 'carry'.

- drive: To **direct** the **move**ment of (a car, truck, bus, etc.)
  →**direct**: To cause (someone or something) to turn, **move**, or point in a particular way

  →**move**: To **go** from one place or position to another

  →**go**: To **move** on a course
- carry: To move (something) while **hold**ing and supporting it
  →**hold**: To **have** or **keep** (something) in your hand, arms, etc.
  →**keep**: To continue **hav**ing or **hold**ing (something)

The conceptual dependency theory by Schank (Schank, 1969) and the preference semantics by Wilks (Wilks, 1972) were examples of the attempts to cope with the above-mentioned problems within the scope of linguistic information understanding, namely, natural language understanding without grounding in the real world. They introduced a small set of conceptual or semantic primitives in order to normalize meaning definitions of words, especially of verbs, so that the system can calculate conceptual or semantic identity or similarity of sentences. That is, they grounded words in symbolic primitives in natural language words or expressions of meta-use. These approaches intended to avoid circular semantic definition and to decrease paraphrastic variety in knowledge representation but the expressive powers were limited because the primitives were given without semantic analysis of a wide-ranged vocabulary, such as the Roget's thesaurus that mental image directed semantic theory is based on, as mentioned in Chapter 8.

On the other hand, the sensory-motor grounded semantics by Roy (Roy, 2005) was a theory for robotic natural language understanding, fully conscious of grounding in a small number of non-symbolic primitives called sensory-motor primitives, directly linked with robotic perception and action. These primitives are defined as combinations of procedural modules, called schemas. They are deemed to be what make conventional black-box procedural definitions of word meanings more analytically transparent. However, this methodology alone is not appropriate for symbol-based semantic computation, such as logical inference based on declarative knowledge representation.

As mentioned above, the formal interpretation of a KRL is conventionally based on the set theory and the intuitive interpretation, exclusively for human users, is given in meta-use of a subset of natural language. It is desirable that any expression in a KRL as a formal language is interpretable in a certain systematic way, both formally and intuitively.

Concerning this point, $\boldsymbol{L}_{md}$ has an advantage over conventional KRLs in the systematic computability of meanings for machine and systematic interpretability for human because an $\boldsymbol{L}_{md}$ expression can be normalized by atomic locus formulas. Especially, for human, the intuitive interpretation of an $\boldsymbol{L}_{md}$ expression is reduced to the (tempo-)logical combination of intuitive interpretations of atomic locus formulas corresponding with language-free loci in attribute spaces while natural language words, especially nouns as Matters and adjectives as Values, are often borrowed for human readability on the way to the complete reduction to atomic locus formulas. It is noticeable that computation on $\boldsymbol{L}_{md}$, as well as on conventional KRLs, is no more than symbol manipulation, but its remarkable distinction exists on its capability of systematic symbol-grounding in the real world (or more exactly, its mental image) and, moreover, it is abstract enough to be applicable to other worlds by replacing the family of the sets of non-logical constants. From the viewpoint of semantic normalization of knowledge representation, the sematic primitives of $\boldsymbol{L}_{md}$ are all the non-logical constants that concern the attribute spaces and loci there.

As for a third problem, a KRL should be provided with a certain systematic interpretation in order to cope with the intrinsic vagueness of the meanings of sensory words, such as 'red' and 'sweet'. The easiest solution of this problem is to include them in the set of semantic primitives besides its original purpose to normalize semantic representation. As the semantic principle of $\boldsymbol{L}_{md}$, it is presumed that all pieces of human knowledge of the physical world are to be reduced to attribute values of matters and their spatiotemporal relations, namely, loci in attribute spaces as mental images. In turn, an attribute space is the model of one of the human sensory fields and assumed to be one kind of tolerance space (Zeeman, 1962) on which sensations are to be mapped with a distance in proportion to their similarity. A tolerance $\tau$ on a set $O$ is a relation $\tau$ that is reflexive and symmetric. A set $O$ together with a tolerance $\tau$ is called a tolerance space (denoted $(O, \tau)$). An attribute value term refers to a region of the attribute space corresponding to a set of similar sensations. Therefore, a sensory word such as 'orange' corresponds to a set of sensations with a certain degree of similarity called 'tolerance' which is located nearer to the 'yellow' region than to the 'green' region in the attribute space of Color ($A_{32}$) whose model can be a color solid. An atomic locus $L(x,y,p,q,a,g,k)$ is to represent a bundle of monotonic paths between regions $p$ and $q$ in the attribute space denoted by $(a,g,k)$, where the term 'monotonic' is not used mathematically but in an intuitive sense, as in "being without significant regions on the way".

# 12

# Multilingual Operation via Mental Image Description Language

Regarding the author's childhood, when his mother tried to make him understand a tale better, she perhaps told it while *paraphrasing* it in a more basic subset of Japanese, based on her *understanding* of the original story. It is thought that if a computer program could take her place, it would be an extreme wonder for ordinary people and a great honor for its creators who should be specialists in natural language processing or, more specifically, natural language understanding. This chapter describes a methodology for multilingual paraphrasing through mental image description language ($L_{md}$) as the most essential working phase of our natural language understanding systems.

As already mentioned, natural language processing has been concerned particularly with how to program computers to process and analyze large amounts of natural language data. However, historically, natural language processing started with machine translation, which necessarily requires natural language generation as well as analysis, which is also the case for the natural language understanding this volume concerns. In natural language processing systems, in the original sense, the languages for input and output are either the same (i.e., monolingual) or different (i.e., cross-lingual). For example, text paraphrasing and text summarization within one language are examples of the former, and machine translation is the typical example of the latter. This chapter describes multilingual operation via $L_{md}$, focusing on multilingual paraphrasing, where target language texts are generated only from the meaning representations in $L_{md}$ of source language texts, namely, with the grammatical descriptions of source language texts detached. This methodology is directly applicable to free translation in distinction from literal translation (Yokota et al., 1984b; Khummmongkol and Yokota, 2018).

In order for efficient multilingual operation, the system should be provided with algorithms and data common to all the natural languages under consideration, separately from those specific to each of them. Based on investigation of several languages, especially, English and Japanese, the author assumes that dependency grammar is commonly applicable for grammatical description and processing of all the natural languages as well as $\boldsymbol{L}_{md}$ representation for semantic description and processing, while morphological description and processing, for example, can be greatly different language by language.

The following sections describe the interaction between dependency grammar and $\boldsymbol{L}_{md}$ mediated by word (or phrase) meaning definitions: how to define the concept and unification parts of a word meaning, how to determine dependency rules in order to optimize the unification parts, and how to apply them to analysis or synthesis of texts in the flow of multilingual paraphrasing.

## 12.1 Meaning Definition

In mental image directed semantic theory, a word meaning is defined by a pair of concept part ($\boldsymbol{L}_{md}$ expression) and unification part (procedure). A word necessarily with a unification part is called a functional word, and otherwise non-functional word. For example, a verb '($x$) eat ($y$)' or an adjective 'red ($x$)' is functional and a noun 'dog' is non-functional. The meaning definitions of functional words in each language are the most important to synthesizing text meanings based on the concepts of the involved words and their dependency structures. A matter (e.g., physical object), usually referred to by a noun as a non-functional word (e.g., *snow*), is to be conceptualized as a conjunction of the mental images of itself and its relations with others that in turn are to be reduced to certain loci in attribute spaces. For example, 'snow is white' and 'snow falls from the sky' (c.f. (5-13), the meaning definition of *snow* in §5.5). That is, a matter concept is a generalized integration of events about all the attributes, implying how the very matter concerns them or not. Intrinsically, every event concept (e.g., *conveyance*) corresponds with the concept part of a certain functional word (e.g., ($x$) *convey* ($y$)) and, therefore, the meanings of functional words are the most essential for natural language understanding.

In general, the steps to define the meanings (i.e., concept and unification parts) of functional words are as follows.

[STEP 1] Determine the set of typical use examples of a functional word $F$, for example, assume $F$ an English verb, *give*, and its basic sentence patterns $\{n_1\ F\ n_2\ n_3,\ n_1\ F\ n_2\ to\ n_p\}$, where $n_i$ is a noun phrase (more exactly, its head) and *to* is the English preposition.

[STEP 2] Synthesize the set of dependency structures of the examples whose roots are the word $F$, for example, $\{F(n_1 n_2 n_3), F(n_1 n_2\, to(n_p))\}$. $F$ is the main governor of the structure and $n_i$ is its $i$-th dependent referred to as *Dep.i*. The preposition *to* is one of dependents of $F$ and simultaneously the governor of the noun $n_p$ referred to as *Dep(to)*.

[STEP 3] Formulate the mental image evoked by the word $F$ in $\boldsymbol{L}_{md}$, for example, $L(x,z,x,y,a,g,k) \wedge x \neq y \wedge y \neq z \wedge z \neq x$, which is the concept part of the meaning of the word. It is noticeable that there is no golden rule to formulate concept parts (i.e., mental images) uniquely but they can be provided in the best way for the applications.

[STEP 4] Determine the mapping rule between the dependency structure and the $\boldsymbol{L}_{md}$ expression, for example, $\{Dep.1 \leftrightarrow x, Dep.2 \leftrightarrow y, Dep.3 \leftrightarrow z\}$ or $\{Dep.1 \leftrightarrow x, Dep.2 \leftrightarrow z, Dep(to) \leftrightarrow y\}$, which is the unification part of the meaning of the word. The notation $Z_1 \leftrightarrow Z_2$ corresponds with the unification command $ARG(Z_1,Z_2)$ introduced in 11.3 which is available for bidirectional mapping between dependency and meaning structures. As well as concept parts, unification parts can be determined according to the applications.

The word meaning definition is often denoted in such a simplified form as (12-1) or (12-2) (c.f., Chapter 14).

$$x\ F\ y\ z : L(x,z,x,y,a,g,k) \wedge x \neq y \wedge y \neq z \wedge z \neq x. \qquad (12\text{-}1)$$

$$x\ F\ z\ to\ y : L(x,z,x,y,a,g,k) \wedge x \neq y \wedge y \neq z \wedge z \neq x. \qquad (12\text{-}2)$$

As mentioned above, word meanings (i.e., concept and unification parts) can be given in the most efficient way for the natural language expressions to be treated in the natural language understanding system. For example, it may be of no problem for a humanoid robot to have the same meaning definition as ‘move’ for ‘walk’, ‘proceed’, ‘go’, or ‘travel’ if it can only walk forward in response to any command using such a verb. This is the very case for our natural language understanding system conversation management system at the present stage (c.f., Chapter 14).

Considering the author’s experiences, it is not very easy to have mental images appropriate for word concepts at once. That is, they should always be hypothetical in order to be improved at any inconvenience. From a quite practical standpoint, word meanings (i.e., concept and unification parts) can be defined so as to be valid within the applications.

## 12.2 Optimization of Grammatical Description for Word Meaning Definition

From the viewpoint of natural language understanding, without a certain mapping algorithm or scheme between surface and meaning structures, any grammar or its products, namely, grammatical descriptions would be of no use. Investigating several languages for natural language understanding so far, the author has come to the conclusion that dependency structure is the best grammatical description to mediate between natural language and $\boldsymbol{L}_{md}$. To his psychological experiences, syntactic dependency as modifier-modifiee relation quite well corresponds with such a human cognitive process that some part of the mental image of the modifiee is instantiated (or made less abstract) by that of the modifier, for example, ‘apple’ versus 'green apple' or 'drink' versus 'drink water', and this is just what the unification parts are destined to simulate. On the other hand, phrase structure grammar is not designed to represent such a relation, but the phrase forming rules help us to group words in a sentence tentatively even if we cannot determine their dependencies at once just as we encounter a certain novel or strange expression, and this is why our natural language understanding system adopts phrase structure grammar as well as dependency grammar. The discussion here is based on the assumption that phrase structure grammar and dependency grammar should work validly for all the human languages in text analysis and generation.

Conventionally, the functionality of dependency grammar has been recognized well and broadly applied to natural language processing to approximate text meaning but there have been no reports except ours about further steps to link dependency structure and mental image due to the lack of its computational models.

In the field of natural language processing, the syntax of a language can be provided so as to match its applications best. For example, the words and grammatical rules can be defined according to the purposes. This is also the case for our natural language understanding system to systematically simulate such a human cognitive process that grammatical description mediates between external and internal representations (i.e, natural language expression and mental image). For example, the dependency grammars of English and Japanese employed in our system are almost the same as the conventionally accepted ones but they are optimized for systematic conversion between dependency structure and $\boldsymbol{L}_{md}$ expression.

Intrinsically, the dependency is a modifier-modifiee relation between two words, but it is quite inefficient to apply dependency rules to texts without any limitations. Actually, it is almost impossible to imagine such dependency rules that are universally applicable, namely, without considering the hierarchy among constituents, where the minimal and the maximal constituents of the hierarchy are a word and a discourse, respectively. Therefore, as already mentioned in Chapter 3, mental image directed semantic theory categorizes dependency rules in four classes, namely, intra-phrase, inter-phrase, inter-clause and inter-discourse (or inter-sentence), in order to limit their application scopes. The rules of each category are applicable only within the same immediately higher constituent (e.g., between phrases within the same clause), representing the dependency relations between the head words of the constituents.

The application of this methodology to English is presented in Chapter 3 and, therefore, consider the case of Japanese here.

Consider such phrasal constructions as '$w_1$, $w_2$, …(,) and $w_n$' and '$w_1$, $w_2$, … (,) or $w_n$'. In mental image directed semantic theory, the semantic interpretations of the clause X involving them are given as (12-3) and (12-4), respectively, where the function $h$ denotes the procedure to convert X to its $\boldsymbol{L}_{md}$ expression, namely, $h$ is the compound of the functions $f$ (text to dependency structure) and $g$ (dependency structure to $\boldsymbol{L}_{md}$) as introduced in Chapter 3. The dependency structures concerning these constructions can be arbitrarily determined as long as they can be systematically converted to the $\boldsymbol{L}_{md}$ expressions (12-3) and (12-4). Such schemes as (12-5) and (12-6) can be appropriate, considering further expanded use (e.g., either $w_1$ or $w_2$), where $[O|\boldsymbol{I}]$ denotes a pair of an operator ($O$) and the list of its operands ($\boldsymbol{I}$).

$$h(X(w_1)) \wedge h(X(w_2)) \wedge \ldots \wedge h(X(w_n)) \tag{12-3}$$

$$h(X(w_1)) \vee h(X(w_2)) \vee \ldots \vee h(X(w_n)) \tag{12-4}$$

$$w_1, w_2, \ldots(,) \text{ and } w_n \Rightarrow \text{and}(w_1\ w_2 \ldots w_n) \Rightarrow [\wedge|\boldsymbol{w}]=[\wedge|\ w_1\ w_2 \ldots w_n] \tag{12-5}$$

$$w_1, w_2, \ldots(,) \text{ or } w_n \Rightarrow \text{or}(w_1\ w_2 \ldots w_n) \Rightarrow [\vee|\boldsymbol{w}]=[\vee|\ w_1\ w_2 \ldots w_n] \tag{12-6}$$

That is, (12-7) and (12-8) are to be defined, where $X([O|\boldsymbol{I}])$ represents $X$ preprocessed in the above ways.

$$h(X([\wedge|\boldsymbol{w}])) \Leftrightarrow h(X(w_1)) \wedge h(X(w_2)) \wedge \ldots \wedge h(X(w_n)) \tag{12-7}$$

$$h(X([\vee|\boldsymbol{w}])) \Leftrightarrow h(X(w_1)) \vee h(X(w_2)) \vee \ldots \vee h(X(w_n)) \tag{12-8}$$

For example, these definitions would make it possible to paraphrase S12-1 as S12-2.

(S12-1) Tom and/or Jim ate beef stake for lunch yesterday.

(S12-2) Tom ate beef stake for lunch yesterday and/or Jim ate beef stake for lunch yesterday.

However, certain further elaboration is required in order to translate such a sentence as (S12-3) into $\boldsymbol{L}_{md}$ because neither Tom nor Jim carried the heavy trunk by himself. In this case, some special operator other than $\wedge$ should be placed at $O$, implying that they are inseparable.

(S12-3) Tom and Jim carried the heavy trunk *together*.

## 12.3 Language Operation via Mental Image Description Language

Figure 12-1 shows the configuration of a multilingual operation system via source language text understanding in $\boldsymbol{L}_{md}$. In principle, multilingual operation system interprets source language text into $\boldsymbol{L}_{md}$ expression in order to understand it and interprets the understanding result into target language text without using any syntactic information of the input. It works as one kind of inter-lingual machine translation system but actually is a multilingual paraphraser as a subsystem of our natural language understanding system.

Multilingual operation system consists of 5 processing units, namely, Source Language Syntax Analyzer, Source Language Meaning Synthesizer, Understanding Processor, Target Language Meaning Synthesizer, and Target Language Text

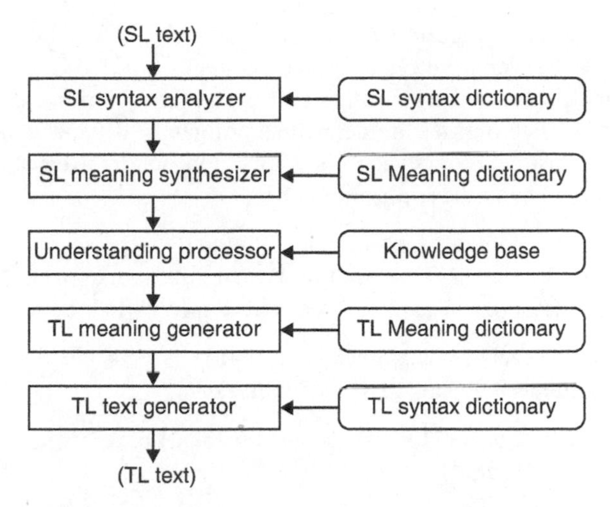


**Fig. 12-1.** Configuration of multilingual operation system.

Generator. For example, consider paraphrasing English text, wholly or partially, in Thai or Japanese by multilingual operation system.

This system works through the following processing steps.

[STEP 1] Source language Syntax Analyzer: Converts the input source language text to dependency structure(s) (often multiple due to ambiguity) using Source Language Syntax Dictionary. For example, consider S12-4 whose most plausible dependency structure is determined as (12-9) by Understanding Processor at last.

(S12-4) Tom takes the book from the town to the school.

takes(Tom book(the) from(town (the)) to(school (the))) (12-9)

[STEP 2] Source Language Meaning Synthesizer: Translates dependency structure into $\boldsymbol{L}_{md}$ as the semantic interpretation of the input, consulting source language Meaning Dictionary.

For example, the dependency structure (12-9) is translated into the $\boldsymbol{L}_{md}$ expression (12-10) according to (12-11), the simplified meaning definition of 'take'. Note that (12-10) is the simplified form of (12-12).

$$\mathrm{L}(Tom,Tom,town,school,\Lambda_t)\Pi\mathrm{L}(Tom,book,Tom,Tom,\Lambda_t), \quad (12\text{-}10)$$

where $\Lambda_t=(A_{12},G_t,k)$.

$$x \text{ take } y \text{ from } p \text{ to } q : \mathrm{L}(x,x,p,q,\Lambda_t)\Pi\mathrm{L}(x,y,x,x,\Lambda_t) \quad (12\text{-}11)$$

$$(\exists\ldots)\ \mathrm{L}(Tom,Tom,z_1,z_2,\Lambda_t)\Pi\mathrm{L}(Tom,z_3,Tom,Tom,\Lambda_t)\wedge\mathrm{town}(z_1)\wedge\mathrm{school}(z_2)\wedge\mathrm{book}(z_3) \quad (12\text{-}12)$$

By the way, as detailed in Chapter 11, 'from *p*' and 'to *q*' can be optional dependents of the verb 'take', to be semantically unified according to the unification part of the meaning definition of each preposition. Here, for the sake of simplicity, the two phrases are defined as obligatory dependents of the verb. In general, the distinction between obligatory and optional dependents can be determined according to the applications.

[STEP 3] Understanding Processor: Evaluates the $L_{md}$ expression(s) (as plausible or not) by reasoning based on knowledge about various matters and postulates (c.f., Chapter 7) registered in Knowledge Base.

For example, such an entailment as (12-13) is possible by applying the postulate of matter as value to (12-10) (see 7.1.1).

$$\mathrm{L}(Tom,Tom,town,school,\Lambda_t)\ \Pi\ \mathrm{L}(Tom,book,Tom,Tom,\Lambda_t) \supset_0 \mathrm{L}(Tom,book,town,school,\Lambda_t) \quad (12\text{-}13)$$

That is, S12-4 entails 'Tom moves the book from the town to the school' and, therefore, (12-10) is equivalent to (12-14).

$$\underline{\mathrm{L}(Tom,Tom,town,school,\Lambda_t)\ \Pi\ \mathrm{L}(Tom,book,Tom,Tom,\Lambda_t)}\ \Pi\ \mathrm{L}(Tom,book,town,school,\Lambda_t) \quad (12\text{-}14)$$

[STEP 4] Target Language Meaning Generator: Searches the $L_{md}$ expression(s) (of the English input) for any Thai (or Japanese) event concept (e.g., verb concept) registered in Target Language Meaning Dictionary and converts it to dependency structure.

For example, the underlined part of (12-14), namely, (12-10) can be converted to Thai and Japanese dependency structures (12-15) and (12-16), respectively (see Table 12-1 for Japanese).

Khleuoen-thi(Tom Kap(Nang-seu) Jak(Meuang) Pai(Rong-rien)) (12-15)

と同時に((持つ(が(トム) を(本))) (移動する(が(トム) から(町) まで(学校)))) (12-16)

Usually, multiple event concepts are to be extracted, here, 'Khleuoen-thi(=move)' and 'Kap(=with)' in Thai, or '移動する(=move)' and '持つ(=hold)' in Japanese, denoted as (12-17)–(12-20), respectively. Moreover, for Japanese here, such a phrase as (12-21) is employed to conjoin the two clauses involving the verbs '持つ' and '移動する'.

*x* Khleuoen-thi Jak *p* Pai *q* : $\mathrm{L}(x,x,p,q,\Lambda_t)$ (12-17)

*x* (Verb) Kap *y* : $\mathrm{L}(x,y,x,x,\Lambda_t)$ (12-18)

*x*が *p*から *q*まで 移動する: $\mathrm{L}(x,x,p,q,\Lambda_t)$ (12-19)

*x*が *y*を 持つ: $\mathrm{L}(x,y,x,x,\Lambda_t) \wedge \mathrm{human}(x)$ (12-20)

**Table 12-1.** Look-up table for Japanese words or phrases.

| Japanese word [pronunciation] | Part of speech | English counterpart |
|---|---|---|
| トム [tomu] | Noun | Tom |
| 本 [hon] | Noun | book |
| 町 [machi] | Noun | town |
| 学校 [gakko] | Noun | school |
| 移動する [idousuru] | Verb | move |
| [motsu] | Verb | hold |
| が [ga] | *Joshi* | NA (subject-indicator) |
| を [wo] | *Joshi* | NA (object-indicator) |
| から [kara] | *Joshi* | from |
| まで [made] | *Joshi* | to |
| と同時に [to-douji-ni] | Conjunction | and simultaneously |

*X*と同時に*Y* : $X \Pi Y$ (*X* and *Y* are clauses) (12-21)

In the meaning definitions of Japanese words, the dependent of *Joshi*, *P* is denoted as *Dep(P)*, for example, *x* and *y* in (12-20) are *Dep(を)* and *Dep(で)*, respectively.

[STEP 5] Target Language Text Generator: Linearizes each dependency structure as a Thai text by employing Target Language Syntax Dictionary.

For example, (12-15) is verbalized as S12-5 in Thai, and (12-16) as S12-6 in Japanese.

(S12-5) Tom Kap Nang-seu Khleuoen-thi Jak Meuang Pai Rong-rien.

(=Tom with the book moves from Town to School.)

(S12-6) トムが本を持つと同時にトムが町から学校まで移動する。

(=Tom holds the book and simultaneously Tom moves from Town to School.)

## 12.4 Question Answering Through Mental Image Description Language

As mentioned above, 5 conceptual categories are employed as terms in an atomic locus formula, namely, Matter, Attribute, Value, Pattern, and Standard. A *Wh* question is assumed to ask for some constant in each place. For example, 'where' and 'when' asks for the Values of the Attributes, Physical Location ($A_{12}$) and Time ($A_{34}$), respectively, and 'what', for some constant Matter. On the other hand, a *Yes/No* question requires some truth value including 'Unknown'. The $\boldsymbol{L}_{md}$ expression for each case is formalized as (12-22) and (12-23), respectively. A *Wh* question is allowed to have multiple interrogatives denoted as $?\boldsymbol{\omega} = \{?\omega_1, ?\omega_2, \ldots, ?\omega_n\}$ in (8) where $?\omega_i$ corresponds with an interrogative.

D(?ω) (12-22)

?D (12-23)

For example, S12-7 and S12-8 are to be interpreted as (12-24) and (12-25), respectively.

(S12-7) When and what did Tom move from Tokyo to Osaka?

(S12-8) Did Tom go to Paris from Tokyo in 2015?

$L(Tom,?x,Tokyo,Osaka,A_{12},G_t,CTY)\Pi$
$L(\_,\_,?p,?q,A_{34},G_t,ST)\Pi L(\_,\_,Past,Past,A_{34},G_t,TD)$ (12-24)

$?L(Tom,Tom,Tokyo,Paris,A_{12},G_t,CTY)\Pi$
$L(\_,\_,+2015years,+2015years,A_{34},G_t,ST)$
$\Pi L(\_,\_,Past,Past,A_{34},G_t,TD)$ (12-25)

In the interpretations above, the granularity or scale of 'Physical Location ($A_{12}$)' or 'Time ($A_{34}$)' is controlled by the Standard parameter as follows.

CTY: Granularity at the unit of 'city',

TD: Granularity at the unit of time division as Past, Present or Future,

ST: Scale of some standard time.

# 13 Computational Model of Japanese for Natural Language Understanding

From the viewpoint of natural language understanding, this chapter focuses on Japanese as a language that is greatly different from English which is the main target of our natural language understanding systems. In mental image directed semantic theory, syntax rules are determined so as to be the fittest for bidirectional translation of dependency structure and mental image description language (i.e., $L_{md}$) expression. For example, the Thai language has the basic structure of clauses (or sentences), Subject + Verb + Object (i.e., linguistic typology as SVO) as well as English and it can be modeled basically by dependency grammar (Tongchim et al., 2008; Khummongkol and Yokota, 2018). On the other hand, Japanese has essentially distinctive structures in comparison with them but it has a good affinity to dependency structure, called *Kakari-uke* (dependent-governor relation), which has been traditionally accepted as the most essential feature of Japanese sentences. Phrase structure grammar as well as dependency grammar is to be defined for Japanese in order to restrict the scope of a dependency rule.

The syntax rules of Japanese presented in the following sections are rather simple compared with actual ones but it is considered that they are sufficient enough for human-robot interaction at the present stage. They are exclusively to process expressions concerning the physical world. The Japanese expressions and their English counterparts cited in the following sections cannot be very good translations of each other because of the small contexts.

## 13.1 Brief Description of Basic Japanese

In Japanese, the verb phrase as a predicate is normally placed at the end of the clause, and grammatical roles such as subject and object are allowed to appear free of order in dependency relations to the verb called *Kakari-uke* (dependent-governor relation) in Japanese. They are indicated by particles called *Joshi* or more exactly *Kaku-joshi* (case-marking *Joshi*) such as *wa*(は), *ga*(が), *kara*(から), *made*(まで), *wo*(を), *ni*(に), and *de*(で) which can be listed as Table 13-1. (N.B., *wa*(は) belongs to *Fuku-joshi* (adverbial *Joshi*), one of the subcategories of *Joshi* in a stricter Japanese grammar.)

*Kaku-joshi* is placed immediately after a noun and works like an English preposition. For example, consider the Japanese sentence, S13-1 (see Table 13-2 for the Japanese words). Its dependency structure is given as (13-1) which is also the case for its syntactic variants in word order, S13-2–5.

(S13-1) トムが大きな箱を家から学校に自転車でとてもゆっくり運ぶ (tomu ***ga*** ookina hako ***wo*** ie ***kara*** gakkou ***ni*** jitensha ***de*** totemo yukkuri hakobu. =Tom carries the big box from his home to the school by bicycle very slowly.)

運ぶ(が(トム)を(箱(大きな))から(家)に(学校)で(自転車)ゆっくり(とても)) (13-1)

(S13-2) 大きな箱をトムが家から学校に自転車でとてもゆっくり運ぶ。

(S13-3) 家から学校に大きな箱を自転車でトムがとてもゆっくり運ぶ。

(S13-4) 自転車で家から学校にトムが大きな箱をとてもゆっくり運ぶ。

(S13-5) とてもゆっくり自転車で家から学校にトムが大きな箱を運ぶ。

**Table 13-1.** List of major *Kaku-joshi*.

| Joshi | Marker for | Usage (English counterpart) |
|---|---|---|
| は [wa] | topic | 象は鼻が長い。(As for an elephant, the trunk is long.) |
| が [ga] | subject | 象は鼻が長い。(As for an elephant, the trunk is long.) |
| を [wo] | direct object | トムが本を買う。(Tom buys a book.) |
| へ [e] | direction | トムは西へ(に)移動する。(Tom moves to the west.) |
| に [ni] | direction<br>goal<br>indirect object<br>time<br>place | トムは東京に(へ)移動する。(Tom moves to Tokyo.)<br>トムが社長に出世する。(Tom is promoted to the president.)<br>トムがメアリに本を与える。(Tom gives Mary the book.)<br>トムが５時に来る。(Tom comes at five o'clock.)<br>トムが公園にいる。(Tom is in the park.) |
| と [to] | participant | トムがメアリと移動する。 (Tom moves with Mary.) |
| で [de] | means | トムがバスで移動する。(Tom moves by bus.) |
| から [kara] | source | トムが東京から移動する。(Tom moves from Tokyo.) |
| まで [made] | limit | トムが北極まで行く。(Tom goes as far as the north pole.) |
| の [no] | various relations | トムの家(Tom's home), 机の引き出し(the drawer of the desk) |

**Table 13-2.** Look-up table for Japanese words.

| Japanese word [pronunciation] | Part of speech | English counterparts |
|---|---|---|
| 象 [zou] | Noun | elephant |
| 鼻 [hana] | Noun | trunk, nose |
| 長い [nagai] | Adjective | long |
| 本 [hon] | Noun | book |
| 買う [kau] | Verb | buy |
| トム [tomu] | Noun | Tom |
| 西 [nishi] | Noun | west |
| 移動する [idousuru] | Verb | move |
| 東京 [toukyou] | Noun | Tokyo |
| 社長 [shachou] | Noun | president |
| 出世する [shusse-suru] | Verb | be promoted |
| メアリ [meari] | Noun | Mary |
| 与える [ataeru] | Verb | give |
| …時 [ji] | Noun | … o'clock |
| 来る [kuru] | Verb | come |
| 公園 [kouen] | Noun | park |
| 運ぶ [hakobu] | Verb | carry |
| です [desu] | Verb | be (as a copula) |
| いる [iru] | Verb | be (for existence or state) |
| 箱 [hako] | Noun | box |
| 家 [ie] | Noun | home |
| 北極 [hokkyoku] | Noun | the north pole |
| 行く [iku] | Verb | go |
| 机 [tsukue] | Noun | desk |
| 引き出し [hikidashi] | Noun | drawer |
| 学校 [gakko] | Noun | school |
| 自転車 [jitensha] | Noun | bicycle |
| 花 [hana] | Noun | flower |
| 模様 [moyou] | Noun | pattern |
| 水 [mizu] | Noun | water |
| 少女 [shoujo] | Noun | girl |
| 夕日 [yuuhi] | Noun | sunset |
| 船 [fune] | Noun | ship |
| 窓 [mado] | Noun | window |
| 旅行 [ryokou] | Noun | travel |

*Table 13.2 contd. ...*

*...Table 13.2 contd.*

| Japanese word [pronunciation] | Part of speech | English counterparts |
|---|---|---|
| する [suru] | Verb | do |
| それ [sore] | Pronoun | it |
| こと [koto] | Noun | matter, event |
| 美しい [utsukushii] | Adjective | beautiful |
| 大きな [ookina] | Adjective | big |
| 立派だ [rippada] | Attributive verb | great |
| 巨大だ [kyodaida] | Attributive verb | huge |
| 寝る [neru] | Verb | go to bed; lie down |
| 売る [uru) | Verb | sell |
| とても [totemo] | Adverb | very |
| ゆっくり [yukkuri] | Adverb | slowly |
| 測る [hakaru] | Verb | measure |
| れる [reru] | Auxiliary verb | can (be able to) … |
| ない [nai] | Auxiliary verb | not |
| および [oyobi] | Conjunction | and |
| そして [soshite] | Conjunction | and |
| しかし[shikashi] | Conjunction | but |
| が [ga] | *Joshi* | NA (subject marker) |
| を [wo] | *Joshi* | NA (object marker) |
| から [kara] | *Joshi* | from |
| に [ni] | *Joshi* | to, in, at, … |
| で [de] | *Joshi* | by, with, in, … |
| *な [na] | *Joshi* | for exclamation (偉大だ<u>な</u>(Great!)) |
| | *Joshi* | for prohibition (走る<u>な</u>(Don't run.)) |
| *か [ka] | *Joshi* | for question or so |
| *よ [yo] | *Joshi* | for compulsive confirmation or so |
| *ね [ne] | *Joshi* | for sympathetic confirmation or so |
| て[te] | *Joshi* | and |
| ば [ba] | *Joshi* | for hypothetical condition |
| 、 | *Kuten* | comma (,), colon (:), or so |
| 。 | *Touten* | the ending mark of every kind of sentence |
| 「…」 | *Hyouki-fugo* | quotation marks such as '…' and "…" |

**Shu-joshi* (one of subcategories of *Joshi*), placed at sentence ends, specifically to express special intention or emotion

## 13.2 Phrase Structure Grammar for Japanese

JPR-1 to JPR-6 are examples of Japanese phrase structure grammar rules, where the special symbols * and + represent an integer ≥ 0 and an integer≥1, denoting the repetition times, respectively. In JPR-2 for a noun phrase, the adjective phrase involved is optional, which is denoted by the pair of parentheses. (See Table 13-3 for abbreviations of parts of speech and phrases of Japanese.)

(JPR-1) AJP⇔ Ad*Aj$^+$ (e.g., とても 美しい =very beautiful)

(JPR-2) NP⇔ Ad*Aj$^+$N$^+$ (J) (e.g., とても 美しい 花 模様 (を) = very beautiful flower pattern)

(JPR-3) YP⇔ Y$^+$Ax*(J) (e.g., 測れる (よ) = can measure, (you know); 美しい (か)=be beautiful (?); 美しく 立派だ (な)=beautiful and great (!))

(JPR-5) ADP⇔Ad$^+$ (e.g., とても ゆっくり = very slowly)

(JPR-6) COP⇔Co (e.g., および, しかし = and, but)

**Table 13-3.** List of parts of speech and phrases of Japanese, and their abbreviations.

| Part of speech (abbreviation) | Phrase (abbreviation) |
|---|---|
| Adjective (Aj) | Adjective phrase (AJP) |
| Adverb (Ad) | Adverb phrase (ADP) |
| Auxiliary verb (Ax) | Noun phrase (NP) |
| Conjunctive (Co) | Verb phrase (VP) |
| Noun (N) (including Pronoun) | Attributive verb phrase (VaP) |
| Joshi (J) | Yougen phrase (YP) (AJP, VP or VaP) |
| Verb (V) | Conjunctive phrase (COP) |
| Attributive Verb (Va) | |
| †Yougen (Y) (Aj, V or Va) | |

†Yougen include *Doushi* (verbs), *Keiyoushi* (adjectives) and *Keiyou-doushi* (attributive verbs) in Japanese.

## 13.3 Dependency Grammar for Japanese

Dependency is called *Kakari-uke* (dependent-governor relation) in Japanese. JDR-1 to JDR-13 are examples of Japanese dependency grammar rules.

(JDR-1) N←J (e.g., 花 が)

(JDR-2) N←N (e.g., 花 模様)

(JDR-3) Aj←N (e.g., 美しい 模様)

(JDR-4) Ad←Aj (e.g., とても 美しい)

(JDR-5) Ad←Ad (e.g., とても ゆっくり)

(JDR-6) Y←Ax (e.g., 行かれる)

| | | |
|---|---|---|
| (JDR-7) | NP←NP | (e.g., 花 模様) |
| (JDR-8) | NP←YP | (e.g., 学校 に 行く = go to school) |
| (JDR-9) | YP→NP | (e.g., (トム が) 行く 学校 = the school which Tom goes to) |
| (JDR-10) | ADP←YP | (e.g., とても ゆっくり 移動する = move very slowly) |
| (JDR-11) | AJP←YP | (e.g., とても 美しい　です = be very beautiful) |
| (JDR-12) | YP←COP | (e.g., 行く そして (買う) = go and (buy)) |
| (JDR-13) | COP→YP | (e.g., (行く) そして 買う = (go) and buy) |

In these rules, *K←U* and *U→K* mean that the head *U* (*uke*) governs the dependent *K* (*kakari*) forward and backward, respectively, and both the cases are to be formulated as (13-2) for computation. A head word (*U*) such as verb appearing in an natural language expression often takes multiple dependent words ($K_1 K_2 \ldots K_n$). Therefore, this should be more generally formulated as (13-3).

$$U(K) \tag{13-2}$$

$$U(K_1 K_2 \ldots K_n) \quad (n \geq 1) \tag{13-3}$$

The rules JDR-1 to JDR-11 are valid within a clause, while JDR-12 and JDR-13 are for inter-clause dependency. Among intra-clause rules, JDR-1 to JDR-6 are for intra-phrase and JDR-7 to JDR-11 are for inter-phrase. The intra-phrase rule JDR-1, for example, reads that Joshi (J) can be a head of a noun (N) only within its own noun phrase (NP) defined by JPR-2. On the other hand, each of JDR-7 to JDR-11 for inter-phrase dependencies actually denote the relation between the head words in either phrase, which is also the case for the inter-clause rules, JDR-12 and JDR-13.

In general, a dependency structure of a text is to be given as a nesting set of the units in the form of (13-3). For example, the grammatical description (i.e., dependency structure) of S13-6 is given as (13-4) by employing the dependency rules above.

(S13-6) トムがとても美しい少女ととても速くドライブした。
(tomu ***ga*** totemo utsukushii shoujo ***to*** totemo hayaku doraibu-shita.
=Tom drove with the very beautiful girl very fast.)

ドライブした((が)トム (と)少女(美しい(とても)) 速く(とても))　　(13-4)

## 13.4 Sentence and Discourse of Japanese

According to the set of dependency rules above, the root of a dependency grammar (i.e., *U* in (13-3)) should be a Yougen for a simple sentence (or a clause) (dependency grammar condition-1) or a conjunction for a compound or complex sentence (dependency grammar condition-2). In other words, a sentence is defined as a grammatical constituent that satisfies either of these conditions in dependency grammar. This definition of a sentence for Japanese is all the same as that for English, as presented in 3.2.3, except for *Yougen*.

A discourse in Japanese, as well as in English, can be a maximal grammatical constituent which consists of a series of sentences combined with discourse connectives, such as 'ところで(*tokorode*=by the way)' and '以下の通り(*ikano-toori*=as follows)', and conjunctions, such as 'および(*oyobi*=and)', 'そして(*soshite*=and)', 'または(*matawa*=or)', and 'なぜならば(*nazenaraba*=because)'. A discourse connective should be the root of the dependency structure of the discourse.

Again, in the author's opinion, it is almost impossible to formalize all allowable expressions within a single grammar, such as phrase structure grammar or dependency grammar. Therefore, a natural language processing system should be provided with a certain mechanism to evaluate semantic plausibility of natural language expressions largely filtered through certain grammars, which is the case for our natural language understanding system based on mental image directed semantic theory. For example, a Japanese sentence containing multiple verbs without conjunctions is approximately treated as a set of multiple clauses combined with the conjunction 'および(*oyobi*=and)' or 'そして(*soshite*=and)' in our systems at the present stage, as detailed in 13.6.

## 13.5 Sentence Types of Japanese and Phrasing

In 3.5, Table 3-2 shows the orthodox correspondences between the types of English sentences and the kinds of tasks for robots to execute. This is not always the case for Japanese because there are no clear distinctions among sentences in Japanese. That is, in Japanese, the tasks are to be designated by special words or inflectional forms of predicative words called *Yougen* (such as verbs and adjectives) which conjugate as *mizen-kei* (irreal form), *renyou-kei* (continuative form), *shuushi-kei* (predicative form), *rentai-kei* (attributive form), *katei-kei* (hypothetical form), and *meirei-kei* (imperative form). For example, Table 13-4 shows the conjugations of verb '走る(*hashiru*=run)' and adjective '速い(*hayai*=fast)') whose inflectional forms are embedded in the places marked by @. Table 13-5 shows the relations between conjugations of *Yougen* and the sentence types (i.e., declarative, interrogative, and imperative).

As mentioned in Chapter 3, viewed from pragmatics, every sentence type (i.e., declarative, interrogative, or imperative) can be employed for another intention, like a rhetorical question. For example, we do not need to answer such a question as 'そんなもの誰が分かるか。(sonna mono dare ga wakaru ka. = Who knows such a thing?)' when we are asked so because its answer should be '誰も分からない。(dare mo wakaranai. = Nobody knows.)' For another example, when people utter '赤信号だ。(akashingou da. = Red signal!)' at a crossing, they surely

**Table 13-4.** Conjugations of Yougen: Verb '走る(*hashiru*=run)', adjective '速い(*hayai*=fast)' and attributive verb '立派だ(*rippada*=great)'

| **Inflectional forms <example contexts>** | **@** | **Example usages** | **English counterparts** |
|---|---|---|---|
| Mizen-kei | | | |
| <Xが@ない。> | V | トムが走らない。 (tomu ga hashiranai) | Tom does not run. |
| | Aj | トムが速くない。 (tomu ga hayakunai) | Tom is not fast. |
| <Xが@う。> | Va | トムが立派だろう。 (tomu ga rippadarou) | Tom will be great. |
| Renyou-kei | | | |
| <Xが@YP。> | V | トムが走り歩く。 (tomu ga hashiri aruku) | Tom runs and walks. |
| | Aj | トムが速く歩く。 (tomu ga hayaku aruku) | Tom runs fast. |
| | Va | トムが立派に走る。 (tomu ga rippani hashiru) | Tom runs greatly. |
| <Xが@た。> | Va | トムが立派だった。 (tomu ga rippadatta) | Tom was great. |
| <Xが@ない。> | Va | トムが立派でない。 (tomu ga rippadenai) | Tom is not great. |
| Shuushi-kei | | | |
| <Xが@。> | V | トムが走る。 (tomu ga hashiru) | Tom runs. |
| | Aj | トムが速い。 (tomu ga hayai) | Tom is fast. |
| | Va | トムが立派だ。 (tomu ga rippada) | Tom is great. |
| <Xが@か。> | V | トムが走るか。 (tomu ga hashiru ka) | Does Tom run? |
| | Aj | トムが速いか。 (tomu ga hayai ka) | Is Tom fast? |
| | Va | トムが立派か。 (tomu ga rippa ka) | Is Tom great? |
| <Xが@とVP。> | V | トムが走ると言う。 (tomu ga hashiru to iu) | Tom says that he runs. |
| | Aj | トムが速いと聞く。 (tomu ga hayai to kiku) | I hear that Tom is fast. |
| | Va | トムが立派だと知る。 (tomu ga rippada to shiru) | I know that Tom is great. |

*Table 13-4 contd. ...*

*...Table 13-4 contd.*

| **Inflectional forms <example contexts>** | **@** | **Example usages** | **English counterparts** |
|---|---|---|---|
| Rentai-kei | | | |
| <Xが@NP...> | V | トムが走る道路…<br>(tomu ga hashiru douro…) | the road where Tom runs |
| | Aj | トムが速い場合…<br>(tomu ga hayai baai…) | the case where Tom is fast |
| | Va | トムが立派な態度…<br>(tomu ga rippana taido…) | Tom's great attitude |
| Katei-kei | | | |
| <Xが@ば、...> | V | トムが走れば、…<br>(tomu ga hashire ba…) | If Tom runs, … |
| | Aj | トムが速ければ、…<br>(tomu ga hayakere ba…) | If Tom is fast, … |
| | Va | トムが立派ならば、…<br>(tomu ga rippanara ba…) | If Tom is great, … |
| Meirei-kei | | | |
| <X@。> | V | 速く走れ。<br>(hayaku hashire) | Run fast! |
| <X@よ。> | V | 速く走れよ。<br>(hayaku hashire yo) | Run fast, you know! |

**Table 13-5.** Sentence types and phrasing in simple sentences.

| **Sentence type** | **Phrasing in a simple sentence [Examples]** |
|---|---|
| Declarative | Affirmative expression:<br>…predicative form 。[トムが走る。(tomu ga hashiru. = Tom runs.); メアリが美しい。(meari ga utsukushii. = Mary is beautiful.)] |
| | Negative expression for verbs and adjectives:<br>…irreal form ない 。[トムが走らない。(tomu ga hashira nai. = Tom does not run.)] |
| | Negative expression for attributive adjectives:<br>…continuative formない。[トムが立派でない。(tomu ga rippade nai. = Tom is not great.)] |
| Interrogative | …declarative か 。 [トムが走るか。(tomu ga hashiru ka. = Does Tom run?); トムが走らないか。(tomu ga hashira nai ka. = Does Tom not run?)] |
| | …***interrogative word***…declarative か 。<br>[どこでトムが走るか。(***doko*** de tomu ga hashiru ka. = Where does Tom run?)] |

*Table 13-5 contd. ...*

*...Table 13-5 contd.*

| **Sentence type** | **Phrasing in a simple sentence [Examples]** |
|---|---|
| Imperative | …imperative form (よ)。[走れ(よ)。(hashire (yo). = Run(, you know)!)] |
| | …continuative form なさい(よ)。[寝なさい(よ)。(ne nasai (yo). = Go to bed (, you know)!] |
| | …continuative form てください(よ)。[寝てください。(ne te kudasai (yo). = Please, go to bed (, you know)!] |

imply a special intention of caution. This unorthodox usage of a declarative can be treated as omission of a certain expression such as '道路を横断するな。(douro wo oudansuru na. = Do not cross the road.)' or '止まれ。(tomare. = Stop!)'.

## 13.6 Syntax and Semantics of Japanese Discourse

Viewed from the standpoint of natural language understanding, syntactic and semantic description of a Japanese discourse are almost all the same as in English. That is, a Japanese discourse can be defined as any unit of connected speech or writing longer than a sentence and there are two types of words or phrases to characterize it, namely, discourse connectives and discourse markers (Yokota, 1999).

In Japanese, as well as in English, a discourse connective, such as *sarani* (更に = moreover), *soshite* (そして = and), *shikashinagara* (しかしながら = however), *mo* (も(*Fuku-joshi*) = too) or *ikano-toori* (以下の通り = as follows) has a function to express the semantic connection between the discourse units prior to and following it. On the other hand, a Japanese discourse marker, such as *oo* (おお = oh), *ee* (ええ = oh), *sate* (さて = well, by the way), *mitaina* (みたいな = like), *sorede* (それで = so), *souyatte* (そうやって = so), *naruhodo* (なるほど = I see, okay), *tsumari* (つまり = I mean, in other words), *yo* (よ(*Shu-joshi*) = you know) or *owakarideshou-ga* (お分かりでしょうが = you know) is rather syntactically independent with a pragmatic function or role to manage the flow of conversation without changing any semantic content (i.e., meaning) of the discourse. Therefore, here, discourse markers are excluded from grammatical or semantic description of a discourse.

All the same as the case of English described in Chapter 3, mental image directed semantic theory defines a Japanese discourse as a series of sentences (or utterances) by (13-5), where a discourse $D$ is a series of sentences $S$, respectively containing a discourse connective $C_d$ and denoted as $S[C_d]$. In turn, the discourse connective $C_d$ is defined by (13-6) as a list of words or phrases, not limited to single conjunctions (i.e., $Co$), where the special symbol $\phi$ corresponds with the case of its omission or implicit establishment of relation (mostly of the conjunction *and*) which is often observed in actual scenes, especially in writing, like parataxis.

$D \Leftrightarrow S^n[C_d] \Leftrightarrow S[C_d]\, S[C_d] \ldots S[C_d]$ (n≥1) (13-5)

$C_d \Leftrightarrow \phi|Co|$そして(soshite)|更に(sarani)|しかしながら(shikashinagara)|
以下の通り(ikanotoori)|も(mo)|... (13-6)

The first sentence appearerd in the form of discourse (13-5), as well as the following ones, can contain a discourse connective that is linked to some prior undescribed or unspoken context. For example, when you find your friend back from jogging, you can give her such an utterance as "そうやって、君は健康管理しようとしてるんだ。(**souyatte**, kimi wa kenkoukanri-shiyou to shiteirunda. = *So* you're trying to keep fit)" or "僕も健康管理に注意しないといけない。(boku **mo** kenkoukanri ni chuuishinai to ikenai. = I should pay some attention to my fitness, *too*"). Moreover, as easily understood, the definition of a discourse holds also for a single sentence consisting of multiple clauses.

Consider a discourse $D_0$, formulated as (13-7), consisting of two units $D_1$ and $D_2$ and a discourse connective $C_d$.

$D_0 = D_1\, C_d\, D_2$ (13-7)

In all the same way as the processing of a single sentence described in 3.2.4, $D_0$ is to be converted into the dependency structure (13-8) and then into the semantic representation in $\boldsymbol{L}_{md}$ (13-9). The processes to translate natural language into dependency structure and dependency structure into $\boldsymbol{L}_{md}$ are denoted as functions $f$ and $\mathcal{g}$, respectively, and $f(D_i)$ and $\mathcal{g}(f(D_i))$ represent the dependency structure and the $\boldsymbol{L}_{md}$ expression of $D_i$, respectively.

The root (i.e., head word) of the dependency structure of a discourse should be a discourse connective, and there are, therefore, two inter-discourse (or inter-sentence) dependency rules, namely, (JDR-14) and (JDR-15), where $H_w$ denotes the head word of the dependent discourse unit. When the dependent discourse unit is a simple sentence, $H_w$ is Yougen, and otherwise a discourse connective.

$\mathfrak{f}(D) = C_d(\mathfrak{f}(D_1)\ \mathfrak{f}(D_2))$ (13-8)

$\mathcal{g}(\mathfrak{f}(D)) = \mathcal{g}(C_d(\mathfrak{f}(D_1)\ \mathfrak{f}(D_2)))$ (13-9)

(JDR-14) $H_w \leftarrow C_d$

(JDR-15) $C_d \rightarrow H_w$

On the other hand, a Japanese sentence containing multiple clauses without conjunctions is approximately treated as a set of multiple clauses connected with the conjunction (or discourse connective) 'そして(*soshite*=and)' in mental image directed semantic theory.

For example, S13-7 consists of four clauses with predicative verb phrases ('歩く(*aruku* = *walk*)' and '見つけた(*mitsuketa* = found)') and adjectives ('美しい(*utsukushii* = beautiful)' and '大きい(*ookii* = big)') embedded. That is, here, a sentence with multiple clauses is recognized and processed as a discourse or its equivalent. This sentence is decomposed into a set of simple sentences conjoined by multiple use of conjunction 'そして(soshite)' to be paraphrased as S13-7'. The underlined nouns shared by the neighboring clauses in S13-7 or common to the neighboring sentences in S13-7' refer to the same matters, namely, to which the same matter terms are to be assigned in the corresponding $\boldsymbol{L}_{md}$ expressions. Such nouns are put immediately after *Yougen* of *rentai-kei* (attributive form), respectively functioning both as a relative and as its antecedent in English. Japanese has no relatives, infinitives or so in English but its syntax is powerful enough to express the contents conveyed by English sentences or clauses of every construction, where *Yougen* play the central roles by conjugation.

(S13-7) 夕日が美しい海辺を歩くトムが見つけたカメは大きい。(yuuhi ga utsukushii umibe wo aruku tomu ga mituketa kame wa ookii. = The turtle is big which Tom found when he walked on the seashore where the sunset was beautiful.)

(S13-7') カメは大きい。そして、トムがカメを見つけた。そして、トムが海辺を歩く。そして、海辺は夕日が美しい。

The process of this type of paraphrasing is digested as follows.

[STEP 1] Partitioning between *rentai-kei* and a noun as in (13-10).

夕日が美しい/海辺を歩く/トムが見つけた/カメは大きい (13-10)

[STEP 2] Conversion of each division to dependency structure as in (13-11), where $J_1$–$J_3$ are Joshi to be required and supplemented to form a simple sentence.

{大きい(は(カメ)), 見つけた(が(トム) $J_1$(カメ)), 歩く($J_2$(トム)を(海辺)), 美しい($J_3$(海辺)が(夕日))} (13-11)

[STEP 3] Estimation of Joshi to be supplemented as (13-12) according to the information fed back from Meaning Synthesizer and Understanding Process following Syntactic Analyzer of Multilingual Operation System, see Chapter 12.

{$J_1$=を, $J_2$=が, $J_3$=は} (13-12)

For another example, consider S13-8 to be decomposed as (13-13). Its remarkable feature is that the first division contains two subjects (marked by が) but actually the noun phrase 'トムが' is the subject of the verb phrase '会った', that is, S13-8 is construed as the sentence S13-9 with the rest, modifying the indirect object 'ジム', embedded in. Therefore, S13-8 can be paraphrased as S13-10.

(S13-8) トムが　メアリが買った本を読んだジムに昨日会った。(tomu ga meari ga katta hon wo yonda jimu ni kinou atta. = Yesterday, Tom saw Jim who read the book which Mary bought.)

トムが メアリが買った/本を読んだ/ジムに昨日会った (13-13)

(S13-9) トムがジムに昨日会った。(tomu ga jimu ni kinou atta. = Yesterday, Tom saw Jim.)

(S13-10) メアリが買った本を読んだジムにトムが昨日会った。

The paraphrasing schemes above are formulated in (i)–(iii) as follows, where *X* and *Y* are a clause or its equivalent. The notation *A*⇒*B*⇒*C* implies translation from natural language text (*A*) into dependency structure of the paraphrase (*B*) and into $\boldsymbol{L}_{md}$ expression (*C*).

i) Construction of *Yougen* of *rentai-kei*

This construction is formulated as $Xy_t\ n_s\ Y$, where $y_t$ is the *rentai-kei* of a *Yougen y* and $n_s$ represents the noun shared by the neighboring clauses $Xy_tn_s$ and $n_sY$, respectively. The scheme for paraphrasing it is given as (13-14), where $X'(n_s)\ y_e$ and $Y'(n_s)$ represent the paraphrases of $Xy_tn_s$ and $n_sY$ with $n_s$ embedded in properly, respectively. The *rentai-kei* of *y* is replaced with its *shuushi-kei* $y_e$.

X $y_t$ $n_s$ Y⇒そして( f(X'($n_s$)$y_e$) f(Y'($n_s$)))⇒g ( f(X'($n_s$) $y_e$))∧g ( f(Y'($n_s$)) (13-14)

For example, S13-11 is paraphrased as S13-11' and translated into the dependency structure (13-15).

(S13-11) 窓が巨大な船でメアリが旅行する(mado ga kyodaina fune de meari ga ryokousuru. = Mary travels by the ship whose windows are huge.)'

(S13-11') 船は窓が巨大だ。そして、メアリが船で旅行する。

そして(巨大だ((は)船(が)窓)) 旅行する(が(メアリ)で(船)) (13-15)

In this example,

{*X*= '窓が巨大な', *Y*= 'でメアリが旅行する', $y_t$ = '巨大な', $n_s$= '船'}
and
{$X'(n_s)y_e$ = '船は窓が巨大だ', $Y'(n_s)$ = 'メアリが船で旅行する', $y_t$ = '巨大だ'}.

ii) Construction of *Yougen* of *renyou-kei* in continuative use

This construction can be generalized as $X\ y_r\ Y$, where $y_r$ is *renyou-kei* of *Yougen* placed between two clauses $Xy_r$ and *Y*. The paraphrasing scheme is formulated as (13-16), where and $y_e$ is *shuushi-kei* of $y_r$.

X $y_r$ Y⇒そして( f(X $y_e$) f(Y)))⇒g ( f (X $y_e$)) ∧g ( f(Y))) (13-16)

For example, S13-12 is paraphrased as S13-12' and converted dependency structure (13-17).

(S13-12) トムが車を買い、メアリがそれで旅行する(tomu ga kuruma wo kai, meari ga sore de ryokou suru. = Tom buys the car, and Mary travels by it.)'

(S13-12') トムが車を買う。そして、メアリがそれで旅行する is converted into the dependency structure (2-28).

そして(買う(が(トム)を(車)) 旅行する(が(メアリ)で(それ))) (13-17)

In this case, $y_r$ = '買い(kai)' and $y_e$= '買う(kau)'.

iii) Construction of verbs of *renyou-kei* in noun use

Among *Yougen*, only verbs of *renyou-kei* can work as a gerund or an infinitive in the noun use of English. This construction can be generalized as $X(Yy_r)$, where $y_r$ is *renyou-kei* of *Yougen* placed after the clause *Y* embedded in the clause *X*. According to this scheme, the verbal construction $Yy_r$ is to be converted to the clausal construction $Y'y_e$. The pair of marks '「…」' is called *Hyouki-kigou*, indicating the boundary of $Y'y_e$ here, used to form an expression equivalent to a 'that-clause' in English.

The paraphrasing scheme is formulated as (13-18), where and $y_e$ is *shuushi-kei* of $y_r$.

X($Yy_r$) ⇒そして(f(X($x_e$)) f($x_e$は「Y'$y_e$」と等しい)

⇒𝒢(f(X($x_e$))) ∧ $x_e$=𝒢(f(Y'$y_e$)) (13-18)

For example, S13-13 should be paraphrased as S13-13'.

(S13-13) トムのメアリとの踊りは美しい。(tomu no meari to no odori wa utsukushii. = Tom's dancing with Mary is beautiful.)

(S13-13') $x_e$は美しい。そして、$x_e$は「トムがメアリと踊る」と等しい。(= $x_e$ is beautiful. And, $x_e$ is equal to 'Tom dances with Mary'.)

It is a matter of course that the paraphrases above can need further elaboration. For example, the paraphrases described above cannot conserve the clausal or verbal subordination in the original expressions. However, they can be appropriate enough for human-robot interaction through natural language in the physical world because the ostensive 4D scenes being described or referred to are the same. That is, the (non-syntactic) difference here between an original expression and its paraphrase is not semantic but pragmatic.

In mental image directed semantic theory, every discourse connective is to be explicitly given a meaning definition with a set of operation commands about how to connect the discourse units and Yokota (Yokota, 1999) has proposed a computational model of the mechanism to select a discourse connective based on the semantic contents of (prior and following) discourse units and the speaker's (or writer's) belief (broadly including knowledge).

Consider the facts being described by sentence $S_1$ and sentence $S_2$ below.

$S_1$: 雨が降っている。(ame ga futte iru. = It is raining.)

$S_2$: トムが家にいる。(tomu ga ie ni iru. = Tom is staying at home.)

When the speaker observes the facts and has such a belief (piece) as $S_3$, she can utter $S_4$.

$S_3$: もし雨が降っているならば、トムが家にいる。(moshi ame ga futte iru naraba, tomu ga ie ni iru. = If it is raining, then he is staying at home.) ($=S_1 \supset S_2$)

$S_4$: 雨が降っているので、トムが家にいる。(ame ga futte iru node, tomu ga ie ni iru. = Tom is staying at home *because* it is raining.)

Her thinking process to make the utterance $S_4$ can be formalized as follows.

Firstly, her observation and belief are given as (13-19) and (13-20), respectively, where $\mathcal{B}$ is the total set of her belief pieces.

$$S_1 \wedge S_2 \tag{13-19}$$

$$S_1 \supset S_2 \in \mathcal{B} \tag{13-20}$$

Secondly, she tries to confirm the belief (13-20) in such a process as (13-21). That is, the belief piece is equivalent to TRUE.

$$S_1 \wedge S_2 \wedge (S_1 \supset S_2) \leftrightarrow S_1 \wedge S_2 \wedge (\sim S_1 \vee S_2) \leftrightarrow S_1 \wedge S_2 \tag{13-21}$$

Finally, she comes to utter $S_4$, concluding that $S_1$ has caused $S_2$.

Therefore, the meaning definition of the conjunctive Joshi 'ので(*node*, *X node Y* = *Y because X*)' can be formalized as (13-22), where the underlined part is out of semantics but in pragmatics because it is specific to the speaker's belief. That is, ので(node) is equal to the logical AND within semantics.

$$X\text{ので}Y \Leftrightarrow X \wedge Y \wedge (\underline{X \supset Y \in \mathcal{B}}) \tag{13-22}$$

In the case of predictive inference based on the fact (or premise) *X*, at least the event referred to by the conclusion *Y* has not been observed yet.

For another example, consider the fact $S_5$ in place of $S_2$.

$S_5$: トムが家にいない。(tomu ga ie ni inai. = Tom is not staying at home.) ($=\sim S_2$)

In this case, she can utter $S_6$ because she comes across contradiction between her observation (13-23) as shown by (13-24) and the conclusion deduced from the fact $S_1$ and her belief $S_3$.

$S_6$: 雨がふっている。しかし、トムが家にいない。(ame ga futteiru. shikashi, tomu ga ie ni inai. = It is raining *but* Tom is not staying at home.)

$$S_1 \wedge S_5 \equiv S_1 \wedge \sim S_2 \tag{13-23}$$

$$S_1 \wedge {\sim} S_2 \wedge (S_1 \supset S_2) \leftrightarrow S_1 \wedge {\sim} S_2 \wedge ({\sim} S_1 \vee S_2) \leftrightarrow \text{FALSE} \tag{13-24}$$

And, therefore, the meaning of しかし(shikashi=*but*) can be defined as (13-25).

$$X \text{ しかし } {\sim} Y \Leftrightarrow X \wedge {\sim} Y \wedge (X \supset Y \in \mathcal{B}) \tag{13-25}$$

As easily understood, it is very important for a robot to discern the semantic part and the pragmatic part in a meaning definition because the latter is usually unobservable, unlike the former (especially concerning the physical world).

It is noticeable that, in general, the material implication statement $p \supset q$ ($\Leftrightarrow {\sim} p \vee q$) does not specify a causal relationship between $p$ and $q$. Therefore, the other two possible cases (13-26) and (13-27) also make (13-20) TRUE, but the utterance $S_7$ and $S_8$ would sound strange and nonsensical, respectively.

$${\sim} S_1 \wedge {\sim} S_2 \tag{13-26}$$

$${\sim} S_1 \wedge S_2 \tag{13-27}$$

$S_7$: 雨が降っていないのでトムが家にいない。(ame ga futteinai node tomuga ie ni inai. = Tom is not staying at home *because* it is not raining.)

$S_8$: 雨が降っていないのでトムが家にいる。(ame ga futte inai node tomu ga ie ni iru. = Tom is staying at home *because* it is not raining.

If $S_7$ sounded good, the belief could be (13-28), implying that *Tom is staying at home if and only if it is raining*.

$$S_1 \equiv S_2 \in \mathcal{B} \tag{13-28}$$

On the other hand, with the same belief, the utterance $S_9$ instead of $S_8$ could be allowed, where the meaning definition of けれども(keredomo = *although*) can be given by (13-29).

$S_9$: (驚いたことには、) 雨が降っていないけれども、トムが家にいる。((odoroita koto niwa,) ame ga futte inai keredomo, tomu ga ie ni iru. = (To my surprise,) Tom is staying at home *although* it is not raining.)

$${\sim} X \text{ けれども } Y \Leftrightarrow {\sim} X \wedge Y \wedge (X \equiv Y \in \mathcal{B}) \tag{13-29}$$

# 14

# Implementation of Mental-Image Based Understanding

We have been developing conversation management system (conventionally, abbreviated as CMS) as the last version of the intelligent system IMAGES in order to simulate intuitive human-robot interaction through natural language. Conversation management system is finally to be provided with a model of human mental-image based understanding according to mental image directed semantic theory. The system at the present stage has been evaluated in comparison with human subjects for our psychological experiment on mental-image based understanding and has shown a good agreement with them in natural language understanding performance.

## 14.1 Configuration of Conversation Management System

Conversation management system is configured, as shown in Fig. 14-1, under further development in Python. *Anna*, the robot the author dreams of, is implemented as artificial intelligence to communicate with a person playing as *Taro* (or so). Anna is a lady-shaped home robot designed for Taro who is a physically handicapped elderly man (see Fig. 1-1). These assumptions are implemented as part of their models. Anna comprehends Taro's intention through dialogue and her final response to his intention is animated graphically. When Anna finds any problem in a situation where she is helping him, she tries to solve the problem by reasoning based on her knowledge and the information acquired by inquiry to the people (Taro and the other residents in the town).

The major modules of conversation management system, as shown in Fig. 14-1, are roughly defined as follows, while it is still under development.

(M1) Natural language understanding module (inside artificial intelligence)

Interprets an input text into mental image description language (i.e., $\boldsymbol{L_{md}}$) expressions and selects the most plausible interpretation by employing all the kinds of knowledge, especially, word-meaning definitions intrinsic to this module.

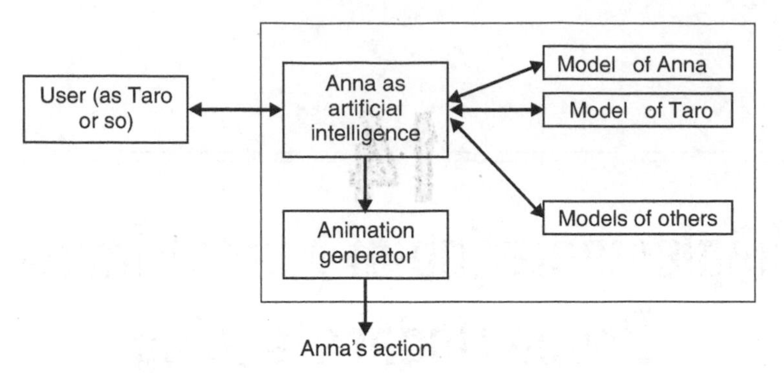


**Fig. 14-1.** Conversation management system.

(M2) Thing-specific model

Consists of knowledge about each person, such as his/her belief, physical mobility, mental tendency, and social activity, or about each non-personal thing (e.g., flower shop, bank, and school), such as its function. Every knowledge piece is represented in $L_{md}$ (e.g., (14-17)–(14-19) below).

(M3) Problem finding/solving module (inside artificial intelligence)

Finds event gaps in $L_{md}$ expressions and cancels them by employing commonsense knowledge pieces, such as postulate of identity of assigned values-type1 ($\mathbf{P}_{V1}$) and postulate of identity of assigned values-type2 ($\mathbf{P}_{V2}$) and the thing-specific models. When problem solving is successful, the solution is sent to the Animation generator in the form of an $L_{md}$ expression. Otherwise, the dialogue partner or so is asked for further information.

(M4) Animation generator

Animates the solution in $L_{md}$ sent from the problem solver.

The details about conversation management system have already been published in (Khummongkol and Yokota, 2016) and, therefore, here is to be focused on mental-image based understanding by conversation management system and its evaluation based on a psychological experiment.

## 14.2 Mental-image Based Understanding versus Conventional Natural Language Understanding

Here, mental-image based understanding by conversation management system is briefly described in comparison with conventional natural language understanding.

Consider such a question as S14-1 to a certain natural language understanding system from its human user. Then, this question can be read roughly as (14-1). It is noticeable that the conjunction in conventional logic is equivalent to SAND (Π) which is denoted as & in (14-1).

(S14-1) When Tom drives with Mary, does she move?

? drive(*Tom*) & with(*Tom,Mary*) → move(*Mary*). (14-1)

For conventional natural language understanding systems to answer such a question correctly, some special piece of knowledge, like (14-2), is required.

drive($x$) & with($x$,$y$) → move($y$). (14-2)

However, people can easily answer 'yes' without employing (14-2). How do they do that? They must employ their mental image evoked by their own experiences. Anna can imitate such a human thinking process. Her understanding of S14-1 can be depicted as Fig. 14-2, where Loc = $(A_{12}, G_t, k)$ and the $\boldsymbol{L}_{md}$ expression reads 'Tom keeps himself in the car, and simultaneously (Π), Tom moves the car from *P* to *Q*, and simultaneously, Tom keeps Mary at his place', and therefore (→) 'Tom moves himself from *P* to *Q*, and simultaneously, Tom keeps Mary at his place' , and therefore, 'Tom moves himself from *P* to *Q*, and simultaneously, Tom moves Mary from *P* to *Q*', and therefore, 'Tom moves Mary from *P* to *Q*', and therefore 'Mary moves from *P* to *Q*, (consequently)'.

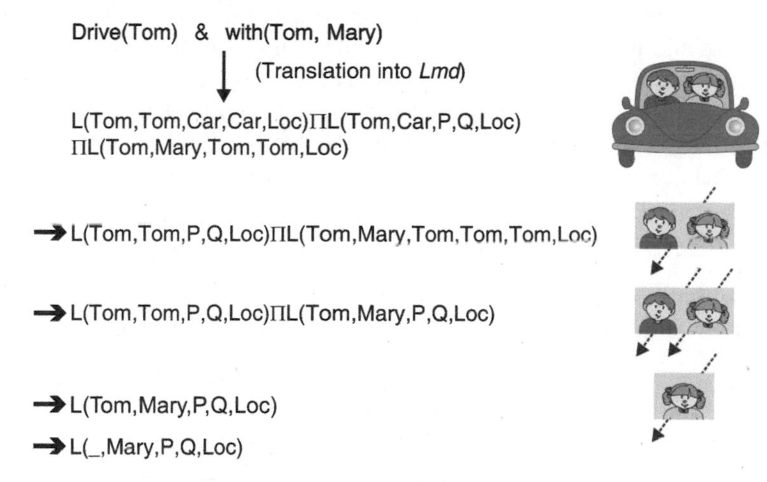

Drive(Tom) & with(Tom, Mary)
(Translation into Lmd)
L(Tom,Tom,Car,Car,Loc)ΠL(Tom,Car,P,Q,Loc)
ΠL(Tom,Mary,Tom,Tom,Loc)
→L(Tom,Tom,P,Q,Loc)ΠL(Tom,Mary,Tom,Tom,Tom,Loc)
→L(Tom,Tom,P,Q,Loc)ΠL(Tom,Mary,P,Q,Loc)
→L(Tom,Mary,P,Q,Loc)
→L(_,Mary,P,Q,Loc)


**Fig. 14-2.** Anna's thinking process for S14-1.

## 14.3 Stimulus Sentences to Conversation Management System and Human Subjects

In order for simple and clear comparison between conversation management system and human subjects, stimulus sentences to them were limited to three types of 4D expressions as follows.

[Type I] Simple sentence + Present particle construction

For example,

(S14-2) Tom was with the book in the bus *running* from Town to University.

[Type II] Simple sentence + Past particle construction

For example,

(S14-3) Tom was with the book in the car *driven* from Town to University by Mary.

[Type III] Simple sentence + Conjunction + Simple sentence

For example,

(S14-4) Tom kept the book in a box *before* he drove the car from Town to University with the box.

As easily convinced, S14-2, 3, and 4 are syntactically ambiguous; that may be rather easy for humans to understand, but it is not the case for robots. For example, consider S14-2. How can the machine know who/what was running from Town to University?—Tom, or book, or bus? Here, to see its syntactic possibilities, Dependency Grammar is employed in order to determine the relations between head words and their dependents. In principle, S14-2 can have twelve possible dependency trees, that is, it can be syntactically ambiguous in twelve ways, as shown in Fig. 14-3, where the arrows labeled *Gij* are directed from governors to dependents. This can be formulated by a set of local dependencies, such as (14-3), where each pair of parentheses is for the alternatives causing the syntactic ambiguity.

{G11, G12, (G13|G13a), (G21|G21a|G21b), G22, (G23|G23a)} (14-3)

According to our psychological experiment, almost all the human subjects reach the most plausible image very easily, like the author's as illustrated in

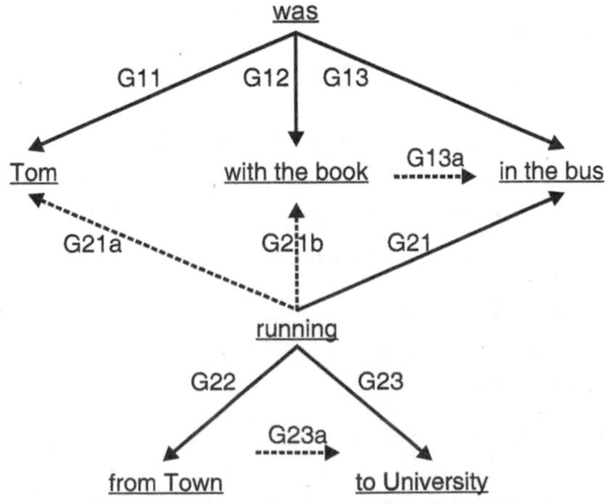


**Fig. 14-3.** Dependencies possible for S14-2.

**Fig. 14-4.** Example of the most plausible images evoked by S14-2.

Fig. 14-4, based on their own daily experiences that correspond directly to the dependency tree defined by (14-4) and can be formulated as (14-5) in $L_{md}$, where the semantic description of the past tense is abbreviated for the sake of simplicity. Figure 14-5 shows the mental image actually sketched by one of the human subjects involved in our psychological experiment.

Hereafter, for the sake of simplicity, simplified $L_{md}$ expressions are adapted. For example, $\Lambda_t = (A_{12}, G_t, k)$, $\Lambda_s = (A_{12}, G_s, k)$, and explicit indications of the quantifiers (i.e., $\forall$ and $\exists$) are omitted, and, moreover, different variables are to refer to different matter or value constants, except anonymous variables (_), without risk of confusion.

$$\{G11, G12, G13, G21, G22, G23\} \quad (14\text{-}4)$$

$$L(Tom, Book, Tom, Tom, \Lambda_t) \Pi L(\_, Tom, Bus, Bus, \Lambda_t) \Pi L(Bus, Bus, Town, Univ., \Lambda_t) \quad (14\text{-}5)$$

Quite in the same way, the most plausible interpretations of S14-3 and S14-4 are given by (14-6) and (14-7), respectively.

**Fig. 14-5.** Mental image about S14-2 sketched by a human subject.

$$\mathrm{L}(Tom,Book,Tom,Tom,\Lambda_t)\Pi\mathrm{L}(\_,Tom,Car,Car,\Lambda_t)\Pi \\ \mathrm{L}(Mary,Mary,Car,Car,\Lambda_t)\Pi\mathrm{L}(Mary,Car,Town,Univ.,\Lambda_t) \qquad (14\text{-}6)$$

$$\mathrm{L}(Tom,Book,Box,Box,\Lambda_t)\bullet(\mathrm{L}(Tom,Tom,Car,Car,\Lambda_t)\Pi \\ \mathrm{L}(Tom,Car,Town,Univ.,\Lambda_t)\Pi\mathrm{L}(Tom,Box,Tom,Tom,\Lambda_t)) \qquad (14\text{-}7)$$

## 14.4 Mental Image-Based Understanding by Conversation Management System

In conversation management system, natural language and $\boldsymbol{L}_{md}$ are translated into each other via a dependency tree employed in conventional natural language processing for grammatical description. The bidirectional translation between a dependency tree and $\boldsymbol{L}_{md}$ is operated by mapping rules assigned to functional words, such as verbs and prepositions, indicating how their context in natural language should be mapped into or generated from the counterpart in $\boldsymbol{L}_{md}$. Here, the mapping rule of a word $W$ is generalized as (14-8). By the way, the mapping rules correspond with ARG commands for unification parts of word meanings presented in 11.3 and the word concepts (e.g., *with(x,y)*) are based on Yokota's own mental experiences.

Context governed by $W \Leftrightarrow$ Concept of $W$ in $\boldsymbol{L}_{md}$ (14-8)

For example, the mapping rules of 'with', 'in', 'move', 'carry', 'run', 'take', and 'bring back' are given by (14-9)–(14-16), where $p\neq q$ and *in(x,y)* are approximated as *at*(*x*,*y*) (c.f., (8-9) and *Postulate of preposition at* ($\mathbf{P}_{at}$) in 8.3) for the sake of simplicity.

Every semantic interpretation (e.g., (14-5)) of a natural language expression (e.g., S14-2) is generated by unifying the word meanings according to the corresponding dependency tree (e.g., (14-4)). In this process, function words, such as verbs and prepositions, are employed for structuring the locus formulas.

$x$ (Verb) with $y$: with$(x,y) \Leftrightarrow \mathrm{L}(x,y,x,x,\Lambda_t)$ (14-9)

$x$ (Verb) in $y$: in$(x,y) \Leftrightarrow \mathrm{L}(_,x,y,y,\Lambda_t)$ (14-10)

$x$ run from $p$ to $q$: run$(x,p,q) \Leftrightarrow \mathrm{L}(x,x,p,q,\Lambda_t)$ or $\mathrm{L}(x,x,p,q,\Lambda_s)$ (14-11)
(i.e., temporal or spatial change)

$x$ carry $y$ from $p$ to $q$: carry1$(x,y,p,q) \Leftrightarrow \mathrm{L}(_,y,x,x,\Lambda_t)\Pi\mathrm{L}(x,x,p,q,\Lambda_t)$ (14-12)

$x$ carry $y$ from $p$ to $q$: carry2$(x,y,p,q) \Leftrightarrow \mathrm{L}(_,x,p,q,\Lambda_t)\Pi\mathrm{L}(x,y,p,q,\Lambda_t)$ (14-13)

$x$ drive $y$ from $p$ to $q$: drive$(x,y,p,q) \Leftrightarrow \mathrm{L}(x,x,y,y,\Lambda_t)\Pi\mathrm{L}(x,y,p,q,\Lambda_t)\wedge\mathrm{car}(y)$
(14-14)

$x$ move $y$ from $p$ to $q$: move$(x,y,p,q) \Leftrightarrow \mathrm{L}(x,y,p,q,\Lambda_t)$ (14-15)

$x$ keep $y$ in $z$: keep$(x,y) \Leftrightarrow \mathrm{L}(x,y,z,z,\Lambda_t)$ (14-16)

On the other hand, matter (or entity) names, such as 'Tom', 'book' and 'bus', are not functional but utilized for disambiguation in syntactic dependency. Our psychological experiment revealed that the human subjects remembered their own experiences in association with the entity names and that they selected the dependency corresponding to their most familiar experience among all the possibilities. For example, the names in S14-2 made the people remember the images in the way as formulated by (14-17)–(14-19), where A ≈>B reads that A evokes B, and + and – denote whether the image is positive (i.e., probable) or negative (i.e., improbable), respectively.

$$\text{Tom} \approx> \{ +L(\_,\text{Tom},\text{Human},\text{Human},\Theta_t), +L(\text{Tom},\text{Tom},p,q,\Lambda_t),\ldots\} \quad (14\text{-}17)$$

$$\text{Book} \approx> \{-L(\text{Book},\text{Book},p,q,\Lambda_t), +L(\text{Human},\text{Book},\text{Human},\text{Human},\Lambda_t),\ldots\} \quad (14\text{-}18)$$

$$\text{Bus} \approx> \{+L(\text{Bus},\text{Bus},p,q,\Lambda_t), +L(\text{Bus},x,p,q,\Lambda_t),+L(\_,\text{Human},\text{Bus},\text{Bus},\Lambda_t),\ldots\} \quad (14\text{-}19)$$

In (14-17), $\Theta_t$ represents 'Quality (or Category) (i.e., $A_{41}$)' with $g = G_t$, and then $+L(\_,\text{Tom},\text{Human},\text{Human},\Theta_t)$ is interpretable as 'it is positive that Tom is a human'. In the same way, $+L(\text{Tom},\text{Tom},p,q,\Lambda_t)$ as 'it is positive that Tom moves by himself', and $-L(\text{Book},\text{Book},p,q,\Lambda_t)$ as 'it is negative that a book moves by itself'.

These images of the matters are derived from their conceptual definitions in the form (6-8) to be employed for probabilistic inference.

It is sure that the subjects reached the most plausible interpretation (14-5) almost unconsciously by using these evoked images for disambiguation. For example, the image for G13 is more probable than that for G13a because of (14-17) and (14-19); G21b is improbable because of (14-18); and the combination of G13 and G21a results in a somewhat strange image that Tom was running in the bus, and therefore G21a is seldom selected.

Furthermore, as well as disambiguation, question-answering in mental-image based understanding was simulated, which is performed by pattern matching between the locus formulas of an assertion and a question, for example, (14-5) for S14-2 and (14-20) for S14-5. Actually, 'carry' is defined in the two ways as (14-12) and (14-13) but, from now on, only one of them is considered for the sake of simplicity. For example, (14-20) adopts the latter.

(S14-5) Did the bus carry the book from Town to University?

$$?\ L(\_,\text{Bus},\text{Town},\text{Univ.},\Lambda_t)\Pi L(\text{Bus},\text{Book},\text{Town},\text{Univ.},\Lambda_t) \quad (14\text{-}20)$$

If (14-5) includes (14-20) as is, the answer is positive, but this is not the case. That is, direct trial of pattern matching to the locus formulas (14-5)–(14-7) does not always directly lead to the desirable outcomes. Therefore, some of the postulates and inference rules introduced in Chapter 7 must be applied to them. They are Postulate of Matter as Value (i.e., $\mathbf{P}_{MV}$), Postulate of Shortcut in Causal Chain (i.e., $\mathbf{P}_{SC}$) and Postulate of Conservation of Values (i.e., $\mathbf{P}_{CV}$), adapted in conversation management system as follows, representing pieces of people's commonsense knowledge about the 4D world.

**P$_{MV}$.** $(\forall\ldots)\ L(z,x,p,q,\Lambda_t)\Pi L(w,y,x,x,\Lambda_t) \supset_0 L(z,x,p,q,\Lambda_t)\Pi L(w,y,p,q,\Lambda_t)$
**P$_{SC}$.** $(\forall\ldots)\ L(z,x,p,q,\Lambda_t)\Pi L(w,y,x,x,\Lambda_t) \supset_0 L(z,x,p,q,\Lambda_t)\Pi L(z,y,p,q,\Lambda_t)$
**P$_{CV}$.** $(\forall\ldots)\ L(z,x,p,p,\Lambda t)\bullet X \supset_0 L(z,x,p,p,t)\bullet(L(z,x,p,p,\Lambda t)\Pi X)$

Postulate of matter as value reads that if '*z* causes *x* to move from *p* to *q* as *w* causes *y* to stay with *x*' then '*w* causes *y* to move from *p* to *q*'. Similarly, postulate of shortcut of causal chain, so that if '*z* causes *x* to move from *p* to *q* as *w* causes *y* to stay with *x*' then '*z* causes *y* to move from *p* to *q* as well as *x*'. Distinguished from these two, postulate of conservation of values is conditional, reading that if '*z* keeps *x* at *p*' then 'it will continue'. That is, it is valid only when *X* does not contradict with $L(z,x,p,p,\Lambda_t)$. The postulates postulate of matter as value and postulate of shortcut of causal chain can be illustrated as Fig. 14-6 and 7, respectively.

On the other hand, as for inference rules, Commutativity Law of Π (**L$_{CS}$**), Elimination Law of SAND (**I$_{ES}$**), Elimination Law of CAND (**I$_{EC}$**), and Substitution Law of locus formulas (**I$_{SL}$**) (c.f., 7.2) are employed, where 'A→B' reads 'A infers B' and 'A↔B' for 'A and B infer each other.'

**I$_{CS}$.** $X\ \Pi\ Y \leftrightarrow Y\ \Pi\ X$
**I$_{ES}$.** $X\ \Pi\ Y \rightarrow X$
**I$_{EC}$.** $X \bullet Y \rightarrow X,\ X \bullet Y \rightarrow Y$
**I$_{SL}$.** $X(\alpha) \wedge \alpha \supset_0 \beta \rightarrow X(\beta)$, where α and β are arbitrary locus formulas.

In order to answer the question S14-5 to S14-2, pattern matching is conducted in order to compare (14-5) and (14-20), as follows. Each step is given by the

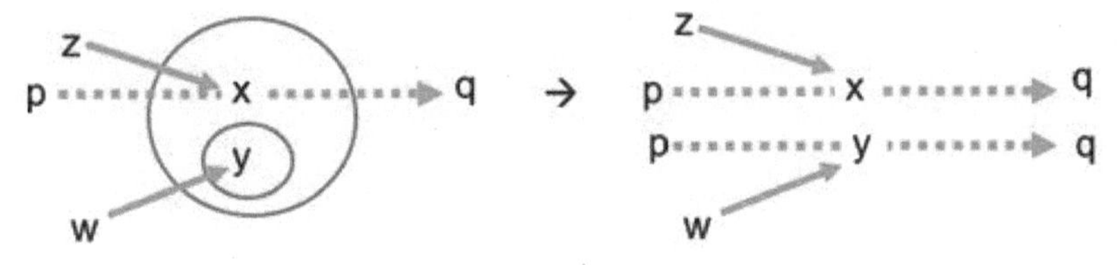


**Fig. 14-6.** Mental image for postulate of matter as value.

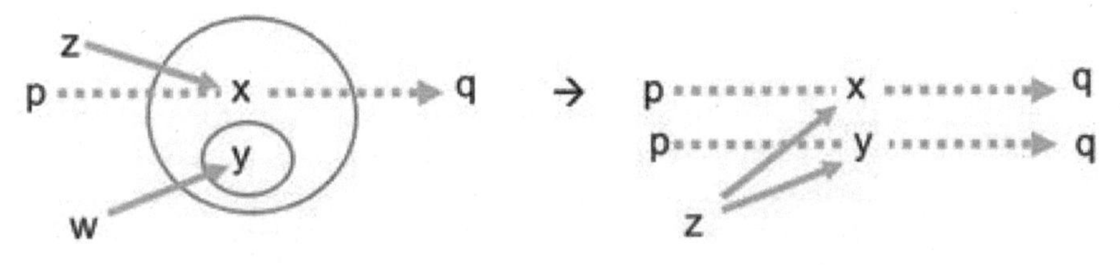


**Fig. 14-7.** Mental image for postulate of shortcut of causal chain.

form 'Σ: Ω' which reads that Ω is deduced from Σ by a certain inference rule of predicate logic.

(14-5), $\mathbf{P}_{MV}$: L(Tom,Book,Bus,Bus,$\Lambda_t$)ΠL(Bus,Bus,Town,Univ.,$\Lambda_t$)
ΠL(_,Tom,Bus,Bus,$\Lambda_t$)

$\mathbf{I}_{CS}$: →L(Bus,Bus,Town,Univ.,$\Lambda_t$)ΠL(_,Tom,Bus,Bus,$\Lambda_t$)
ΠL(Tom,Book,Bus,Bus,$\Lambda_t$)

$\mathbf{I}_{ES}$: →L(Bus,Bus,Town,Univ.,$\Lambda_t$)ΠL(Tom,Book,Bus,Bus,$\Lambda_t$)

$\mathbf{I}_{SC}$: →L(Bus,Bus,Town,Univ.,$\Lambda_t$)ΠL(Bus,Book,Town,Univ.,$\Lambda_t$) Con_1

Apply existential generalization rule of predicate logic (**EG**) and $\mathbf{I}_{SL}$ to (14-5), the definition of '_', and Con_1, then Con_2 below is deduced.

(14-5), Con_1, **EG**, $\mathbf{I}_{SL}$ : L(_,Bus,Town,Univ.,$\Lambda_t$)ΠL(Bus,Book,Town,Univ.,$\Lambda_t$)
Con_2

The pattern matching process finds that (14-20) = Con_2, and then it is proved that the bus carried the book from Town to University.

For another example, consider the stimulus sentence S14-3 and the question S14-6.

(S14-6) Did Mary carry the car from Town to University?

Adopting (14-13) for 'carry', the interpretation of S14-6 can be given by (14-21).

? L(_,Mary,Town,Univ.,$\Lambda_t$)ΠL(Mary,Car,Town,Univ.,$\Lambda_t$) (14-21)

In order to answer the question S14-6 to S14-3, the pattern matching process works as follows.

(14-6), $\mathbf{I}_{CS}$: L(Tom,Book,Tom,Tom,$\Lambda_t$)ΠL(_,Tom,Car,Car,$\Lambda_t$)Π
L(Mary,Car,Town,Univ.,$\Lambda_t$)ΠL(Mary,Mary,Car,Car,$\Lambda_t$)

$\mathbf{P}_{MV}$, $\mathbf{I}_{SL}$: →L(Tom,Book,Tom,Tom,$\Lambda_t$)ΠL(_,Tom,Car,Car,$\Lambda_t$)Π
L(Mary,Car,Town,Univ.,$\Lambda_t$)ΠL(Mary,Mary,Town,Univ.,$\Lambda_t$)

$\mathbf{I}_{CS}$ : →L(Tom,Book,Tom,Tom,$\Lambda_t$)ΠL(_,Tom,Car,Car,$\Lambda_t$)Π Con_3
L(Mary,Mary,Town,Univ.,$\Lambda_t$)ΠL(Mary,Car,Town,Univ.,$\Lambda_t$)

Apply **EG** and $\mathbf{I}_{SL}$ to (14-6), the definition of '_', and Con_3, then Con_4 is deduced.

(14-6), Con_3, **EG**, $\mathbf{I}_{SL}$ : L(Tom,Book,Tom,Tom,$\Lambda_t$)ΠL(_,Tom,Car,Car,$\Lambda_t$)Π

L(_,Mary,Town,Univ.,$\Lambda_t$)ΠL(Mary,Car,Town,Univ.,$\Lambda_t$) Con_4

$\mathbf{I}_{CS}$, $\mathbf{I}_{ES}$: → L(_,Mary,Town,Univ.,$\Lambda_t$)ΠL(Mary,Car,Town,Univ.,$\Lambda_t$) Con_5

Hence, the pattern matching process proves that (14-21) = Con_4 and it is concluded that Mary carried the bus from Town to University.

For the last example, consider the Type III sentence S14-4, and the question S14-7 whose interpretation is given by (14-22).

(S14-7) Did Tom move the book from Town to University?

$$? \ L(\text{Tom,Book,Town,Univ.},\Lambda_t) \tag{14-22}$$

The proof is as follows, where the underlined part is induced by $\mathbf{P}_{CV}$.

$\mathbf{P}_{CV}$, (14-7), $\mathbf{I}_{SL}$: $L(\text{Tom,Book,Box,Box},\Lambda_t)\bullet(\underline{L(\text{Tom,Book,Box,Box},\Lambda_t)}\ \Pi$
$L(\text{Tom,Tom,Car,Car},\Lambda_t)\Pi L(\text{Tom,Box,Tom,Tom},\Lambda_t)\Pi$
$L(\text{Tom,Car,Town,Univ.},\Lambda_t))$

Apply $\mathbf{I}_{CS}$ several times.

$\mathbf{I}_{CS}$: $\rightarrow L(\text{Tom,Book,Box,Box},\Lambda_t) \bullet (L(\text{Tom,Car,Town,Univ.},\Lambda_t)\Pi$
$L(\text{Tom,Tom,Car,Car},\Lambda_t)\Pi L(\text{Tom,Box,Tom,Tom},\Lambda_t)\Pi$
$L(\text{Tom,Book,Box,Box},\Lambda_t))$

$\mathbf{P}_{MV}$, $\mathbf{I}_{SL}$: $\rightarrow L(\text{Tom,Book,Box,Box},\Lambda_t) \bullet (L(\text{Tom,Car,Town,Univ.},\Lambda_t)\Pi$
$L(\text{Tom,Tom,Town,Univ.},\Lambda_t)\Pi L(\text{Tom,Box,Tom,Tom},\Lambda_t)\Pi$
$L(\text{Tom,Book,Box,Box},\Lambda_t))$

Apply postulate of matter as value and substitution law of locus formulas twice.

$\mathbf{P}_{MV}$, $\mathbf{I}_{SL}$ : $\rightarrow L(\text{Tom,Book,Box,Box},\Lambda_t) \bullet (L(\text{Tom,Car,Town,Univ.},\Lambda_t)\Pi$
$L(\text{Tom,Tom,Town,Univ.},\Lambda_t)\Pi L(\text{Tom,Box,Town,Univ.},\Lambda_t)\Pi$
$L(\text{Tom,Book,Town,Univ.},\Lambda_t))$

$\mathbf{I}_{CS}$, $\mathbf{I}_{ES}$ : $\rightarrow L(\text{Tom,Book,Town,Univ.},\Lambda_t)$

In this way, the system finds that (14-22) is deduced from (14-7).

## 14.5 Problem-Finding and Solving in Conversation Management System

This section details conversation management system based on a typical example of dialogue between Taro and Anna. Basically, Anna tries to understand Taro's utterance as a certain request, namely, *intention* for her to do something because her mission, one of Anna's beliefs, is to help him in every aspect. This time, for the sake of simplicity, Taro's utterances were limited to what should represent his intentions explicitly without any rhetorical sophistication, for example, by employing such expressions as "I want…", imperatives, question-word questions and yes-no questions. Under this assumption, Taro's request is logically formulated as (14-23). This formula reads "If *D*, then Taro will get happier (by something *z*, possibly, *D* itself)", where *H-ness* = $(B_{031},G_t,k)$ representing temporal change in *Happiness*, one of subspaces of the attribute *Emotion* ($B_{03}$), and the pair of values *p* and *p*< imply a certain increase in happiness. Hereafter, such a relation *pRq* between two values $p$ and $q$ is to be represented simply by the pair of values $p$ and $pR$, where $R$ is such as $<$ and $\neq$.

$$\begin{aligned}&D\rightarrow(\exists z)L(z,Taro,p,p<,H\text{-}ness)\\&\text{or } D\rightarrow L(\_,Taro,p,p<,H\text{-}ness)\end{aligned} \qquad (14\text{-}23)$$

The aim of Anna's mission is the right hand of (14-23), namely, Taro's getting happier, and she believes that if she realizes $D$, then Taro will get happier. Therefore, she tries to realize $D$, where the general algorithm implemented can be digested as the steps below.

(STEP 1) Evaluate $D$ by the thing-specific models.

(STEP 2) Remove any part of $D$ that is already feasible as is, namely, of no problem to be solved. The remaining part is the goal event ($X_G$) to be attained by Anna.

(STEP 3) Find the current event ($=X_C$) and compose $X_C \bullet X_T \bullet X_G$ as defined in (9-1).

(STEP 4) Assign constants to variables appearing in $X_C$ or $X_G$ by asking User (=Taro, here) or consulting the thing-specific models.

(STEP 5) Solve $X_T$ according to the postulate of *continuity in attribute values* as explained by (9-2) and (9-3).

(STEP 6) Align and revise $X_T$ to be feasible (in animation) by consulting the models concerned.

(STEP 7) If another event gap is detected in association with the revised $X_T$, find another $X_C$ to compose another $X_C \bullet X_T \bullet X_G$ and return to STEP 4. Otherwise, send the solution to Animation Generator.

For example, consider such a situation as follows.

*On the first day when they meet, Anna is with Taro and Taro is thirsty. She is not so familiar with his home yet.*

In this situation, the dialogues between them and the performances of conversation management system are assumed as follows.

*Taro> "I'm thirsty."*

Firstly, Anna's semantic understanding of this utterance, namely, its meaning, is given by (14-24), reading that Taro is thirsty due to $x$.

$$\begin{aligned}&(\exists x)L(x,Taro,+Th,+Th,Vitality)\\&\text{or } L(\_,Taro,+Th,+Th,Vitality)\end{aligned} \qquad (14\text{-}24)$$

According to Table 3-2, her flat response could be as follows, disappointing him, but his intention must be otherwise.

*Anna> "I know."*

*Taro> "Oops!"*

Secondly, instead, she tries to infer Taro's intention, namely, pragmatic understanding, by employing such a piece of knowledge as (14-25) in the model of ordinary people, reading that when Taro is thirsty, he drinks water and then it makes him not thirsty.

L(_,Taro,+Th,+Th,Vitality). $\supset_1$. drink(Taro,Water)
•L(Water,Taro,+Th,-Th,Vitality) (14-25)

Here, the verb concept drink is roughly defined as (14-26). That is, 'x drink y' is defined as 'x moves y from Mouth to Stomach which are with/at/in x'. (Any further elaborated definition is available in $\boldsymbol{L}_{md}$ as necessity in the principle of the quasi-symbolic image system.)

(λx,y)drink(x,y)
⇔(λx,y)L(x,y,Mouth,Stomach,$\Lambda_t$)
ΠL(x,{Mouth,Stomach},x,x,$\Lambda_t$) (14-26)

Finally, her pragmatic understanding is formulated as (14-27), reading if 'Taro drinks water and it makes him not thirsty (=*D*)', then 'he will get happier'.

drink(Taro,Water)•L(Water,Taro,+Th,-Th,Vitality)
→L(_,Taro,p,p<,H-ness) (14-27)

Therefore, their dialogue should be continued as follows.

*Anna> "You want to drink some water, don't you?"*

*Taro> "Yes, right."*

For the next stage, Anna gets started with problem-finding and solving in the part *D* of (14-27), namely, the underlined part and its full expression as a locus formula is given as (14-28).

L(Taro,Water,Mouth,Stomach,$\Lambda_t$)

ΠL(Taro,{Mouth,Stomach},Taro,Taro,$\Lambda_t$)
•L(Water,Taro,+Th,-Th,Vitality) (14-28)

At STEP 1 for problem-finding and solving in the prescription, Anna infers (14-29) from Taro's model.

Taro≈>{+L(Taro,Water,Mouth,Stomach,$\Lambda_t$)

ΠL(Taro,{Mouth,Stomach},Taro,Taro,$\Lambda_t$),
+L(Taro,Water,p,p,$\Lambda_t$)} (14-29)

This is about Taro's probability directly associated with *D* in (14-23), implying "Taro can send water from his mouth to his stomach", and "Taro can keep water somewhere (*p*)".

Then, at STEP 2, she knows water can make Taro not thirsty and finds no problem in Taro's drinking water. Then, she becomes aware of the fact that she has only to make water be at/in his mouth.

At STEP 3, she does not know where (=*p*) Taro keeps water. Therefore, her problem (=$X_T$) is to know its current place (=$X_C$) and make it move to his mouth. The event concerning the location of water is formulated as (14-30) and (14-31). If Taro has water in his mouth currently and Anna knows that (namely, already included in $X_C$), there is no event gap.

$\underline{L(z_1,Water,?p,?p, \Lambda_t)}\bullet X_T$

$\bullet L(Taro,Water,Mouth,Stomach, \Lambda_t)$ (14-30)

$X_T = L(z_2,Water,?p,Mouth, \Lambda_t)$ (14-31)

At STEP 4, Anna asks Taro about the place of the water.

*Anna> "Where is the water?"*

This is the verbalization of the underlined part of (14-30). for problem solving. That is, the semantic definition of "*where be* x?" is "$L(z_1,x,?p,?p, \Lambda_t)$".

*Taro> "In the fridge."*

Anna understands that *?p = Fridge*.

At STEP 5, she tries to solve the underlined part (=$X_T$) of (14-32). Here, for the sake of simplicity, it is deemed that "x *in* y" = "x *at* y" (c.f., (14-10)).

$L(Taro,Water,Fridge,Fridge,\Lambda_t)$

$\bullet \underline{L(z_2,Water,Fridge,Mouth,\Lambda_t)}$ (14-32)

At STEP 6, Anna supposes that $z_2$ = *Anna*, namely, that she moves the water to Taro' mouth by herself and knows that when Anna moves something, she carries it by herself. (She has no telekinesis to move anything remotely.) This fact is formalized as the postulate (14-33) prepared in Anna's model. This is the elaboration process to make the $\boldsymbol{L_{md}}$ expression feasible by the actor, which, in this case, is Anna.

$L(Anna,x,p,p\neq,\Lambda_t)$

$\leftrightarrow L(Anna,x,Anna,Anna,\Lambda_t)\ \Pi L(Anna,Anna,p,p\neq,\Lambda_t)$ (14-33)

Applying this postulate to (14-32), Anna can deduce (14-34) whose right hand is the revised $X_T$ to be the new goal event to be attained.

$L(Anna,Water,Fridge,Mouth,\Lambda_t)$

$\leftrightarrow L(Anna,Water,Anna,Anna,\Lambda_t)\ \Pi L(Anna,Anna,Fridge,Mouth,\Lambda_t)$ (14-34)

At STEP 7, another event gap is found in the revised $X_T$ as $X_{T1}$ in (14-32). Anna knows that currently, Anna is at Taro and this fact is appended to the previous $X_C$ (=L(Taro,Water,Fridge,Fridge,$\Lambda_t$)) as underlined in (14-35), where $X_{T1}$ is another transit event newly aroused in the previous one.

$$(L(Anna,Anna,Taro,Taro,\Lambda_t)$$
$$\Pi\ L(Taro,Water,Fridge,Fridge,\Lambda_t))\bullet X_{T1}$$
$$\bullet(L(Anna,Water,Anna,Anna,\Lambda_t)$$
$$\Pi L(Anna,Anna,Fridge,Mouth,\Lambda_t)) \quad (14\text{-}35)$$

Again, at STEP 3, the new transit event $X_{T1}$ is solved as (14-36).

$$X_{T1} = L(Anna,Anna,Taro,Fridge,\Lambda_t)$$
$$\bullet\ L(Anna,Water,Fridge,Anna,\Lambda_t) \quad (14\text{-}36)$$

Then, the solution of $X_T$ is finally given as (14-37). This is the final form of the solution that Anna can make feasible by herself.

$$X_T = (L(Anna,Anna,Taro,Fridge,\Lambda_t)$$
$$\bullet\ L(Anna,Water,Fridge,Anna,\Lambda_t))$$
$$\bullet\ (L(Anna,Water,Anna,Anna,\Lambda_t)$$
$$\Pi L(Anna,Anna,Fridge,Mouth,\Lambda_t)) \quad (14\text{-}37)$$

The direct verbalization of (14-37) can be as follows.

*Anna> "Sure, I'll go to the fridge, then take the water from the fridge, and carry it to your mouth from the fridge."*

However, this expression is not elegant. For example, 'your mouth' should be 'you'. For such refinement, the postulate Postulate of Matter as Value, presented in Chapter 7 is available. That is, the deduction below is possible.

$$L(Anna,Anna,Fridge,Mouth,\Lambda_t)$$
$$\Pi L(Taro,Mouth,Taro,Taro,\Lambda_t)$$
$$\rightarrow L(Anna,Anna,Fridge,Taro,\Lambda_t)$$

In verbalization of $\boldsymbol{L}_{md}$ expressions, a simple pattern matching operation is performed in order to search for verb concepts such as (14-9)–(14-16). Conversation management system sends the final solution to the Animation Generator as well, to be translated into a procedure for animation.

## 14.6 Awareness Control of Conversation Management System

We have implemented our theory of mental-image based understanding as the core of conversation management system on a personal computer in Python, a high-level programming language, while it is still experimental and evolving. Conversation management system can understand User's assertions and answer the questions where the locus formulas were given in Polish notation, for example, as $\bullet\Pi ABC$ for $(A\Pi B)\bullet C$. In the actual implementation, the theorem proving process was simplified as the pattern matching process programmed to apply all the possible postulates to the locus formula of the assertion in advance and detect any match with the question in the assertion (extended by the postulates) by using the inference rules on the way. During pattern matching, the system is to control its awareness in a top-down way, driven by the pair of attribute carrier and its attribute contained in the question, for example, 'Book' and 'Physical Location ($\Lambda_t$)', which is very efficient compared with conventional pattern matching methods without employing any kind of semantic information.

The gist of artificial or robotic awareness and its control by conversation management system is as follows.

a) Awareness of something $x$ by an artificial intelligence system is defined as existence of $\boldsymbol{L}_{md}$ expressions about $x$ in its working memory ready for certain inferential computation.
b) An $\boldsymbol{L}_{md}$ expression about $x$ represents some external event involving $x$, corresponding with loci of robotic selective attention scanning the very event.
c) Loci of selective attention are modeled as those in attribute spaces corresponding with robotic sensors or actuators.
d) Because of (b) and (c), a cognitive robot provided with a certain artificial intelligence system can be expected to systematically ground its ideas (i.e., $\boldsymbol{L}_{md}$ expressions in the working memory) in its sensations or actuations.
e) Conversation management system shows the potential for people to control robotic awareness systematically by natural language.

# 15

# Robotic Imitation Guided by Natural Language

As already mentioned, natural language can convey the exact intention of the emitter to the receiver due to the syntax and semantics common to its users. This is not necessarily the case for non-linguistic media, such as photographs, while it can contain much more concrete information about a visual scene than linguistic description. Therefore, it is most desirable to employ both linguistic and non-linguistic information media for communication, namely, multimedia communication. This chapter describes robotic imitation aided by verbal suggestions in addition to demonstration (Yokota, 2007) as one of the phases of integrative multimedia understanding by a robot shown in Fig. 1-1.

## 15.1 Trends in Robotic Imitation

Robotic or artificial imitation is one kind of machine learning on human actions and there have been a considerable number of studies on imitation learning from human actions demonstrated without any verbal hint (e.g., Billard, 2000; Alissandrakis et al., 2003; Nakanishi et al., 2004). In this case, it is extremely difficult for a robot to understand which part of human demonstration is significant or not because there are too many things to attend to. That is, where the attention of the observer should be focused when a demonstrator performs an action is an important issue. Whereas, there have been several proposals to control attention mechanisms efficiently in such top-down ways as guided by the prediction based on sensory data and knowledge of goals or tasks (e.g., Wolfe, 1994; Demiris and Khadhouri, 2006; Katz et al., 2017), they are not realistic when a large number of actions must be imitated distinctively with various speeds, directions, trajectories, etc.

The author has been working on integrated multimedia understanding for intuitive human-robot interaction, that is, interaction between non-expert or ordinary people (i.e., laypeople) and home robots, where natural language is the leading information medium for their intuitive communication (Yokota, 2006,

2012). For ordinary people, natural language is the most important because it can convey the exact intention of the sender to the receiver due to the syntax and semantics common to its users, which is not necessarily the case for other mediums, such as picture. Therefore, the author believes that it is most desirable to realize robotic imitation aided by human verbal suggestion where robotic attention to human demonstration is efficiently controllable based on semantic understanding of the suggestion.

For such a purpose, it is essential to develop a systematically computable knowledge representation language as well as representation-free technologies, such as neural networks, for processing unstructured sensory/motory data. This type of language is indispensable to *knowledge-based* processing, such as *understanding* sensory events, *planning* appropriate actions and *knowledgeable* communication with ordinary people in natural language, and, therefore, it needs to have at least a good capability of representing spatiotemporal events that correspond to humans'/robots' sensations and actions in the real world.

Most conventional methods have provided robotic systems with such quasi-natural language expressions as 'move (*Velocity*, *Distance*, *Direction*)', 'find (*Object*, *Shape*, *Color*)' and so on for human instruction or suggestion, uniquely related to computer programs to deploy sensors/motors (Coradeschi and Saffiotti, 2003; Drumwright et al., 2006). In association with robotic imitation intended here, however, these expression schemas are too linguistic or coarse to represent and compute sensory/motory events, except for in special cases (e.g., Katz et al., 2017).

The Mental Image Directed Semantic Theory (Yokota, 2005) has proposed a model of human attention-guided perception, yielding omnisensory images that inevitably reflect certain movements of the focus of attention of the observer scanning certain matters in the world. More analytically, these omnisensory images are associated with spatiotemporal changes (or constancies) in certain attributes of the matters scanned by the focus of attention of the observer and modeled as temporally parameterized "loci in attribute spaces", so called, to be formulated in a formal language $\boldsymbol{L}_{md}$.

The most remarkable feature of $\boldsymbol{L}_{md}$ is its capability of formalizing spatiotemporal matter concepts grounded in human/robotic sensation, while the other similar knowledge representation languages are designed to describe the logical relations among conceptual primitives represented by lexical tokens (Sowa, 2000). Quite distinctively, $\boldsymbol{L}_{md}$ expression implies what and how should be attended to in human action as analogy of human focus of the attention of the observer movement and, thereby, the robotic attention can be controlled in a top-down way.

## 15.2 Analytical Consideration about Robotic Imitation

In the Webster's dictionary, the verb 'imitate' is defined as 'to be or appear like'. According to this definition, robotic imitation can be defined as (15-1) in $\boldsymbol{L}_{md}$, where

$X(\alpha)$ is an event caused by an agent $\alpha$ and expressed in $\boldsymbol{L}_{md}$. This implies that the imitator (*Imi*=robot) realizes an event *X*(*Imi*) with such an idea (or intention) that it should be of the same quality as the event *X*(*Dem*) caused by the demonstrator (*Dem*=human), where $A_{01}$, $B_{02}$, and $A_{41}$ are the attributes, 'World', 'Location of information', and 'Quality', respectively. The underlined part is interpreted as that the imitator makes *X*(*Imi*) be ($p$=$q$) or come ($p \neq q$) in some world.

(Imi) *imitate* (Dem's X):
$(\exists\ldots)\ \underline{L(\text{Imi}, X(\text{Imi}), p, q, A_{01}, G_t, k_2)}\ \Pi\ L(\text{Imi}, \text{Idea}, \text{Imi}, \text{Imi}, B_{02}, G_t, k_1)$
$\wedge\ \text{Idea} = L(\text{Imi}, X(\text{Imi}), X(\text{Dem}), X(\text{Dem}), A_{41}, G_t, k_3)$ (15-1)

Here, assume that the robot is humanoid, namely, with the same physical structure and functionality as a human and, therefore, *X*(*Imi*) is a copy of *X*(*Dem*) except for *Imi* in place of *Dem*. In other words, the human and the robot can communicate with each other based on the common attributes (employed in the event pattern *X*), which is one case of homogeneous communication presented in Chapter 10. Then, the problem is how the robot acquires and aligns *X*(*Dem*) to be feasible for it as *X*(*Imi*). By the way, it is noticeable that the syntax of $\boldsymbol{L}_{md}$ presented in Chapter 3 must be expanded in order to allow such a higher order expression as (15-1).

## 15.3 Top-down Control of Robotic Attention by Mental Image Description Language

For comprehensible communication with humans, robots must understand natural language *semantically* and *pragmatically*, as described in Chapter 3. Here, as shown in Fig. 3-3, semantic understanding means associating symbols to conceptual images of matters (i.e., objects or events), and pragmatic understanding means anchoring symbols to real matters by unifying conceptual images with perceptual images. In association with robotic imitation here, as shown in Fig. 15-1, the former makes a robot abstractly (i.e., conceptually) aware of which matters and attributes involved in the human demonstration should be attended to, and the latter provides the robot with a concrete idea of real matters with real attribute values significant for imitation. More exactly, semantic understanding in $\boldsymbol{L}_{md}$ of human suggestion enables the robot to control its attention mechanism in such a top-down way that focuses the robot's attention on the significant attributes of the significant matters involved in the human demonstration. Successively, in order for pragmatic understanding in $\boldsymbol{L}_{md}$ of the human suggestion, the robot is to select the appropriate sensors corresponding with the suggested attributes and make them run on the suggested matters so as to pattern after the movements of human focus of the attention of the observer implied by the locus formulas yielded in semantic understanding.

***That is to say, in short,*** $\boldsymbol{L}_{md}$ ***expression suggests to the robot what and how should be attended to in the human demonstration.***

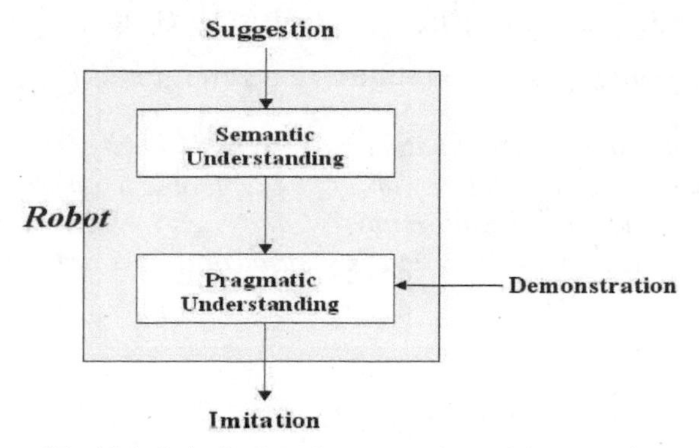


**Fig. 15-1.** Imitation based on suggestion and demonstration.

## 15.4 Theoretical Flow of Robotic Imitation based on Demonstrations and Verbal Hints

As shown in Fig. 15-2, it is assumed that there is a feedback loop between a human and a robot in order for the human to improve his/her previous suggestion/ demonstration and for the robot to correct its previous imitation. Here is described a theoretical flow of the robotic imitation intended by the author, which is almost that of problem finding/solving in the field of artificial intelligence (Yokota, 2006; Khummongkol and Yokota, 2016). Consider the scenario presented below and depicted in Fig. 15-3 and 15-4.

***Scenario:***

***Robby is an intelligent humanoid robot and Tom is its user. Robby is called by Tom and enters his room. This is Robby's first visit there. Robby sees Tom leftward, the yellow table rightward and the green object over the brown pillar forward. After a while, Tom tells Robby "Imitate me to my demonstration and suggestion."***

The sequence of the events assumed to happen is as follows:

[Robby's Perception of the initial situation, $\boldsymbol{S_0}$]

$\boldsymbol{S_0} \leftrightarrow L(\_,O_{21},Brown,Brown,A_{32},G_t,K_{23})$

$\Pi L(\_,O_{22},Green,Green,A_{32},G_t,K_{23})$

$\Pi L(\_,O_{23},Yellow,Yellow,A_{32},G_t,K_{23})$

$\Pi(L(\_,O_{24},Robby,O_{21},A_{12},G_s,K_{21}) \bullet L(\_,O_{24},O_{21},O_{22},A_{12},G_s,K_{21}))$

$\Pi L(\_,O_{24},Fw,Fw,A_{13},G_s, K_{22})$

$\Pi L(\_,O_{25},Robby,Tom,A_{12},G_s, K_{21}) \Pi L(\_,O_{25},Lw,Lw,A_{13},G_s, K_{22})$

$\Pi L(\_,O_{26},Robby,O_{23},A_{12},G_s, K_{21})\Pi L(\_,O_{26},Rw,Rw,A_{13},G_s, K_{22})$

$\wedge pillar(O_{21})\wedge object(O_{22})\wedge table(O_{23})\wedge ISR(O_{24})\wedge ISR(O_{25}) \wedge ISR(O_{26})$

*Robby's recognition of the situation (i.e., the underlined part of the scenario) is still rough due to its economical working mode that is to be specified by each Standard (or precision)* $K_{2n}$. *The attributes* $A_{32}$ *and* $A_{13}$ *are 'Color' and 'Direction', respectively. The values* Fw, Lw *and* Rw *stand for 'forward', 'leftward' and 'rightward', respectively.*

[Tom's Intention_1, ***Int₁***]

$\boldsymbol{Int_1}\leftrightarrow L(\underline{Robby,Robby,Robby,O_{11},A_{12},G_t,K_{11}})\Pi$

$L(Robby,Robby,V_{11},V_{11},A_{16},G_t,K_{12})\Pi$

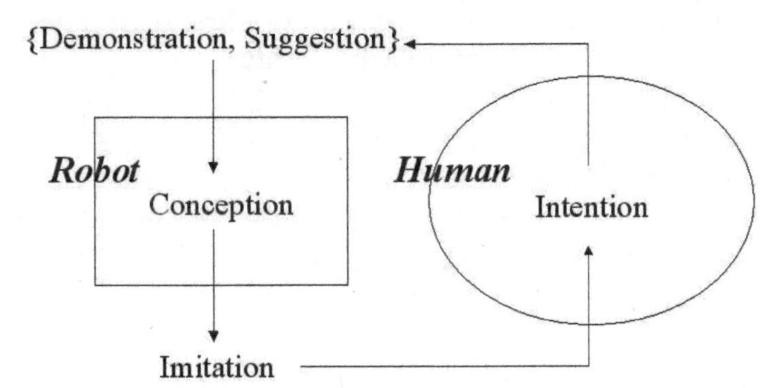


**Fig. 15-2.** Feedback loop between human and robot.

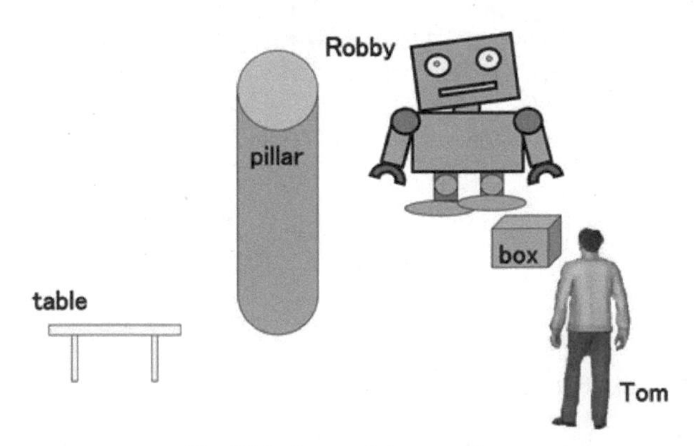


**Fig. 15-3.** Scene of the scenario.

$L(\_,O_{13},Robby,O_{12},A_{12},G_s,K_{11})\Pi$

$L(Robby,O_{13},Dis,Dis,A_{44},G_t,K_{13})$

$\underline{\wedge stool(O_{11})}\wedge pillar(O_{12})\wedge ISR(O_{13})$

*Tom wants Robby to go to the stool without touching the pillar on the way. Tom is conscious that every attribute value to specify Robby's action is essentially vague but should be imitated within certain tolerance associated with each Standard (or precision)* $K_{1i}$. *For example, Tom cannot specify* the value $V_{11}$ *of 'Velocity (A16)' exactly, like 'walk at 2 m/sec', but he believes that he can show it by demonstration. In general, the topology between two distinct objects is formulated as above with the attribute* $A_{44}$ *of the ISR involved (e.g.,* $O_{13}$*). The values* Dis *and* Meet *stand for 'disjoint' and 'meet (or touch)', respectively.*

**<SESSION_1>**

[Tom's Suggestion_1, $\boldsymbol{T_1}$ and Demonstration_1, $\boldsymbol{D_1}$]

$\boldsymbol{Int_1 \Rightarrow T_1, D_1}$

$\boldsymbol{T_1} \leftrightarrow$ "Move to the stool like this."

$\boldsymbol{D_1} \leftrightarrow$ Fig. 15-4

*Tom decides to verbalize only the underlined part of Intention_1,* $\boldsymbol{Int_1}$ *saliently with the belief that the rest can be included in his demonstration. Tom converts (or translates) Intention_i,* $\boldsymbol{Int_i}$ *into Suggestion_i,* $\boldsymbol{T_i}$ *and Demonstration_i,* $\boldsymbol{D_i}$.

[Robby's Semantic_Understanding_1, $\boldsymbol{S_1}$]

$\boldsymbol{T_1 \Rightarrow S_1}$

$\boldsymbol{S_1} \leftrightarrow (\exists x,k_1)L(Robby,Robby,Robby,x,A_{12},G_t,k_1)\wedge stool(x)$

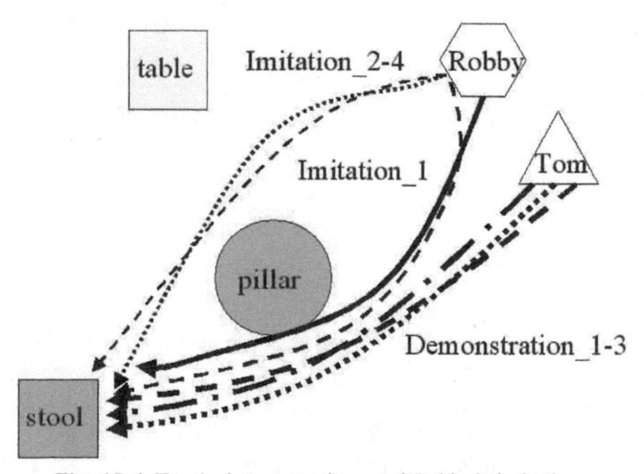


**Fig. 15-4.** Tom's demonstrations and Robby's imitations.

*Robby interprets Suggestion_i,* $\boldsymbol{T_i}$ *into Semantic_ Understanding _i,* $\boldsymbol{S_i}$*. The variable 'x' is not yet anchored to the 'real stool' in the real environment because 'stool' refers to 'seat' without a back but can have various appearances.*

[Robby's Pragmatic_Understanding_1, $\boldsymbol{P_1}$ and Default_1, $\boldsymbol{Def_1}$]

$\boldsymbol{D_1 \Rightarrow PD_1}$

$\boldsymbol{S_1, PD_1 \Rightarrow P_1, Def_1}$

$\boldsymbol{P_1} \leftrightarrow$ L(Robby,Robby,Robby,$O_{22}$,$A_{12}$,$G_t$,$K_{21}$)$\wedge$stool($O_{22}$)

$\boldsymbol{Def_1} \leftrightarrow$ L(Robby,Robby,1m/sec,1m/sec,$A_{16}$,$G_t$,_)$\wedge$...

*The 'Location ($A_{12}$)' is attended to according to Suggestion _1. Tom's Demonstration_1 makes Robby aware that the word 'stool' should be anchored to the 'green object* $O_{22}$*' in the real environment. Robby conceives that it should approach the stool at its standard (or precision)* $K_{21}$*.* $\boldsymbol{PD_i}$ *refers to Robby's perception of Tom's Demonstration_i(i.e., X(Dem) in (15-1)).* $\boldsymbol{Def_i}$ *is inferred from* $\boldsymbol{PD_i}$ *(i.e., X(Imi) in (15-1)) as the default specification for the attributes not explicit in Suggestion_i.*

[Robby's Imitation_1, $\boldsymbol{I_1}$]

$\boldsymbol{P_1, Def_1 \Rightarrow I_1}$

$\boldsymbol{I_1} \leftrightarrow$

Fig. 15-4

*Robby imitates Tom's Demonstration_1 according to Pragmatic _ Understanding_i,* $\boldsymbol{P_i}$ *and Default_i,* $\boldsymbol{Def_i}$*.*

----- Resetting the situation to the initial situation $\boldsymbol{S_0}$-----

**<SESSION_2>**

[Tom's Suggestion_2, $\boldsymbol{T_2}$ and Demonstration_2, $\boldsymbol{D_2}$]

$\boldsymbol{I_1 \Rightarrow PI_1}$

$\boldsymbol{Int_1, \sim PI_1 \Rightarrow Int_2}$

$\boldsymbol{Int_2 \Rightarrow T_2, D_2}$

$\boldsymbol{T_2} \leftrightarrow$"Don't touch the pillar."

$\boldsymbol{D_2} \leftrightarrow$ Fig. 15-4

*Tom perceives Robby's Imitation_1,* $\boldsymbol{I_1}$ *as* $\boldsymbol{PI_1}$*. He denies it and creates Intention_2,* $\boldsymbol{Int_2}$ *followed by Suggestion_2 and Demonstration_2.*

[Robby's Semantic_Understanding_2, $\boldsymbol{S_2}$]

$\boldsymbol{T_2 \Rightarrow S_2}$

$\boldsymbol{S_2} \leftrightarrow \sim(\exists x,y,k_1,k_2)$L(x,y,Robby,$O_{21}$,$A_{12}$,$G_s$,$k_1$)
$\Pi$L(Robby,y,Dis,Meet,$A_{44}$,$G_t$,$k_2$)$\wedge$ISR(y)$\wedge$pillar($O_{21}$)

*Robby becomes aware that its imitation has been denied at the change of attribute 'Topology ($A_{44}$)' from* '**Dis**joint' *to* '**Meet**' *because the existence of the Imaginary Space Region* (y) *between Robby and the pillar is undeniable (c.f.,* $\mathbf{P}_{\mathbf{IS}}$ *(Postulate of Arbitrary Production of Imaginary Space Region) in Chapter 7).*

[Robby's Pragmatic_Understanding_2, $\boldsymbol{P_2}$ and Default_2, $\boldsymbol{Def_2}$]

$\boldsymbol{D_2 \Rightarrow PD_2}$

$\boldsymbol{S_2, P_1, \sim Def_1, PD_2 \Rightarrow P_2, Def_2}$

$\boldsymbol{P_2} \leftrightarrow \boldsymbol{P_1} \wedge \underline{\mathrm{L}(\_,\mathrm{O}_{27},\mathrm{Robby},\mathrm{O}_{12},\mathrm{A}_{12},\mathrm{G}_s,\mathrm{K}_{21})\Pi\mathrm{L}(\mathrm{Robby},\mathrm{O}_{27},\mathrm{Dis},\mathrm{Dis},\mathrm{A}_{44},\mathrm{G}_t,\mathrm{K}_{22}) \wedge \mathrm{pillar}(\mathrm{O}_{21}) \wedge \mathrm{ISR}(\mathrm{O}_{27})}$

$\boldsymbol{Def_2} \leftrightarrow \mathrm{L}(\mathrm{Robby},\mathrm{Robby}, 1\mathrm{m/sec}, 1\mathrm{m/scc},\mathrm{A}_{16},\mathrm{G}_t,\_) \wedge \ldots$

*According to Semantic_ Understanding_2, the 'Location ($A_{12}$)' of Robby and the pillar and their 'Topology ($A_{44}$)' are particularly attended to, and the underlined part is conceived in addition to Pragmatic_Understanding_1. No special attention is paid to the other attributes unmentioned yet.*

[Robby's Imitation_2, $\boldsymbol{I_2}$]

$\boldsymbol{P_2, Def_2 \Rightarrow I_2}$

$\boldsymbol{I_2} \leftrightarrow$ Fig. 15-4

-----Resetting the situation to the initial situation $\boldsymbol{S_0}$-----

**<SESSION_3>**

[Tom's Suggestion_3, $\boldsymbol{T_3}$ and Demonstration_3, $\boldsymbol{D_3}$]

$\boldsymbol{I_2 \Rightarrow PI_2}$

$\boldsymbol{Int_2, \sim PI_2 \Rightarrow Int_3}$

$\boldsymbol{Int_3 \Rightarrow T_3, D_3}$

$\boldsymbol{T_3} \leftrightarrow$"Much closer to the stool like this."

$\boldsymbol{D_3} \leftrightarrow$ Fig. 15-4

[Robby's Semantic_Understanding_3, $\boldsymbol{S_3}$]

$\boldsymbol{T_3 \Rightarrow S_3}$

$\boldsymbol{S_3} \leftrightarrow (\exists x,y,q_{21},q_{22},k_{21},k_{22},l_{21})$
$\mathrm{L}(\mathrm{Robby},\mathrm{Robby},\mathrm{Robby},\mathrm{O}_{22},\mathrm{A}_{12},\mathrm{G}_s,k_{21})$
$\Pi\mathrm{L}(x,y,\mathrm{Robby},\mathrm{O}_{22},\mathrm{A}_{12},\mathrm{G}_s,k_{21})\Pi\mathrm{L}(\mathrm{Robby},y,q_{21},q_{22},\mathrm{A}_{02},\mathrm{G}_t,k_{22})$
$\wedge (q_{22} \ll l_{21}) \wedge \mathrm{ISR}(y) \wedge \mathrm{stool}(\mathrm{O}_{22})$

*Robby becomes aware that some final value* $q_{22}$ *in the Length ($A_{02}$) between Robby and the stool should be much shorter (<<) than some value* $l_{21}$ *intended by Tom.*

[Robby's Pragmatic_Understanding_3, $\boldsymbol{P_3}$ and Default_3, $\boldsymbol{Def_3}$]

$\boldsymbol{D_3 \Rightarrow PD_3}$

$\boldsymbol{S_3, P_2, \sim Def_2, PD_3 \Rightarrow P_3, Def_3}$

$\boldsymbol{P_3} \leftrightarrow \boldsymbol{P_2} \wedge L(\_,Robby,Robby,O_{22},A_{12},G_s,\_)\Pi$
$L(\_,O_{28},Robby,O_{22},A_{12},G_s,\_)\Pi$
$L(Robby,O_{28},10\ m,0.3\ m,A_{02},G_t,\_) \wedge ISR(O_{28}) \wedge stool(O_{22})$

$\boldsymbol{Def_3} \leftrightarrow L(Robby,Robby, 1\ m/sec, 1\ m/sec,A_{16},G_t,\_) \wedge \ldots$

*The 'Length ($_{02}$)' (of the Imaginary Space Region $O_{28}$) between Robby and the stool is attended to at Demonstration_3 and the final value* $q_{22}$ *is estimated as* 0.3 m *from the perception of Demonstration_3,* $\boldsymbol{PD_3}$, *where the value* '10m' *is the initial length between Robby and the stool. No attention is paid to the other attributes unmentioned yet.*

*All the variables are to be instantiated by the real objects.*

[Robby's Imitation_3, $\boldsymbol{I_3}$]

$\boldsymbol{P_3, Def_3 \Rightarrow I_3}$

$\boldsymbol{I_3} \leftrightarrow$ Fig. 15-4

----- Resetting the situation to the initial situation $\boldsymbol{S_0}$-----

**<SESSION_4>**

[Tom's Suggestion_4, $\boldsymbol{T_4}$ and Demonstration_4, $\boldsymbol{D_4}$]

$\boldsymbol{I_3 \Rightarrow PI_3}$

$\boldsymbol{Int_3, \sim PI_3 \Rightarrow Int_4}$

$\boldsymbol{Int_4 \Rightarrow T_4, D_4}$

$\boldsymbol{T_4} \leftrightarrow$ "More speedily."

$\boldsymbol{D_4} \leftrightarrow$ ***Null*** *(No demonstration)*

[Robby's Semantic_Understanding_4, $\boldsymbol{S_4}$]

$\boldsymbol{T_4 \Rightarrow S_4}$
$\boldsymbol{S_4} \leftrightarrow (\exists p,k)L(Robby,Robby,p,p,A_{16},G_t,k) \wedge p>k$

[Robby's Pragmatic_Understanding_4, $\boldsymbol{P_4}$ and Default_4, $\boldsymbol{Def_4}$]

$\boldsymbol{S_4, P_3, \sim Def_3 \Rightarrow P_4, Def_4}$

$\boldsymbol{P_4} \leftrightarrow \boldsymbol{P_3} \wedge L(Robby,Robby, 3\ m/sec, 3\ m/sec,A_{16},G_t,\_)$

$\boldsymbol{Def_2} \leftrightarrow \ldots$

*The 'Velocity ($A_{16}$)' is especially attended to and is increased up to* 3 m/sec. *No special attention is paid to the other attributes unmentioned.*

[Robby's Imitation_4, $\boldsymbol{I_4}$]

$\boldsymbol{P_4, Def_4 \Rightarrow I_4}$

$\boldsymbol{I_4} \leftrightarrow$ Fig. 15-4

----- Resetting the situation to the initial situation $\boldsymbol{S_0}$-----

**<SESSION_5>**

[Tom's Suggestion_5, $\boldsymbol{T_5}$ and Demonstration_5, $\boldsymbol{D_5}$]

$\boldsymbol{I_4 \Rightarrow PI_4}$

$\boldsymbol{Int_4, \sim PI_4 \Rightarrow Int_5}$

$\boldsymbol{Int_5 \Rightarrow T_5, D_5}$

$\boldsymbol{T_5} \leftrightarrow$"A little less speedily."

$\boldsymbol{D_5 \leftrightarrow Null}$

[Robby's Semantic_Understanding_5, $\boldsymbol{S_5}$]

$\boldsymbol{T_5 \Rightarrow S_5}$

$\boldsymbol{S_5} \leftrightarrow (\exists p,k,v_1,v_2,c)L(Robby,Robby,v_2,v_2,A_{16},G_t,k) \wedge v_2=v_1 \times (1-c) \wedge (c>0.0 \wedge c \cong 0.0)$

*Robby becomes aware that the velocity should be changed into some value* $v_2$ *slightly smaller than another value* $v_1$.

[Robby's Pragmatic_Understanding_5, $\boldsymbol{P_5}$ and Default_5, $\boldsymbol{Def_5}$]

$\boldsymbol{S_5, P_4, \sim Def_4 \Rightarrow P_5, Def_5}$

$\boldsymbol{P_5 \leftrightarrow P_4} \wedge L(Robby,Robby,2.4\ m/sec, 2.4\ m/sec,A_{16},G_t,\_)$

$\boldsymbol{Def_5} \leftrightarrow \ldots$

*The 'Velocity ($A_{16}$)' is attended to again. The alternative velocity* $v_2$ *is calculated as* 2.4m/sec *according to Robby's belief that* $v_1$=3m/sec *and* c=0.2.

[Robby's Imitation_5, $\boldsymbol{I_5}$]

$\boldsymbol{P_5, Def_5 \Rightarrow I_5}$

----- Resetting the situation to the initial situation $\boldsymbol{S_0}$-----

**<SESSION_6>**

[Tom's Suggestion_6, $\boldsymbol{T_6}$ and Demonstration_6, $\boldsymbol{D_6}$]

$\boldsymbol{I_5 \Rightarrow PI_5}$

$\boldsymbol{Int_5, \sim PI_5 \Rightarrow Int_6 (\leftrightarrow Null)}$

$\boldsymbol{Int_6 \Rightarrow T_6, D_6}$

$\boldsymbol{T_6} \leftrightarrow$"Alright."

$\boldsymbol{D_6} \leftrightarrow$ ***Null***

*Tom fails to deny the perception of Imitation_5,* $\boldsymbol{PI_5}$ *and comes to have no another intention (*$\boldsymbol{Int_6} \leftrightarrow$ ***Null****). That is, Tom is satisfied by Robby's Imitation_5 and only tells Robby* "Alright."

[Robby's Semantic_Understanding_6, $\boldsymbol{S_6}$]

$\boldsymbol{T_6} \Rightarrow \boldsymbol{S_6}$

$\boldsymbol{S_6} \leftrightarrow (\exists x,y,k)L(x,y,1,1,B_{01},G_t,k) \wedge person(x)$

*Tom becomes aware that something* 'y' *has evaluated by some person* 'x' *as perfect* '1' *at 'Worth (*$B_{01}$*)' with a certain Standard* 'k'.

[Robby's Pragmatic_Understanding_6, $\boldsymbol{P_6}$ and Default_6, $\boldsymbol{Def_6}$]

$\boldsymbol{S_6}, \boldsymbol{I_5} \Rightarrow \boldsymbol{P_6}, \boldsymbol{Def_6}$

$\boldsymbol{P_6} \leftrightarrow L(Tom,\boldsymbol{I_5},1,1,B_{01},G_t,Tom) \wedge person(Tom)$

$\boldsymbol{Def_6} \leftrightarrow L(Robby, \boldsymbol{I_6},/,/,A_{01},G_t,\_)$

*Finally, Robby pragmatically conceives that Tom is satisfied by Imitation_5,* $\boldsymbol{I_5}$ *at Tom's Standard and believes that the next imitation,* $\boldsymbol{I_6}$ *is not needed to take 'World (*$A_{01}$*)'.*

[Robby's Imitation_6, $\boldsymbol{I_6}$]

$\boldsymbol{P_6}, \boldsymbol{Def_6} \Rightarrow \boldsymbol{I_6}$

$\boldsymbol{I_6} \leftrightarrow$ Null

*Finally, no more imitation is performed.*

-----End of all the sessions-----

## 15.5 Discussion and Perspective Remarks

No athletic coach would succeed in making the pupils understand his demonstrations exactly without any verbal suggestion. The key contribution of this chapter is the proposal of a novel idea of robotic imitation driven by semantic representation of human suggestion, where it is hinted in the formal language $\boldsymbol{L_{md}}$ what and how should be attended to in human action as analogy of human focus of the attention of the observer movement and, thereby, the robotic attention can be controlled in a top-down way. Without such a control, a robot is to attend *simultaneously* to tens of attributes (Table 5-2) of *every* matter involved in human action, as shown in Fig. 1-3. This is not realistic, considering the performants of machine learning and robotic vision understanding today. The author has a good perspective for the proposed theory of robotic imitation based on their previous work utilizing $\boldsymbol{L}_{md}$ for robot manipulation by natural language (Yokota, 2006, 2007, 2012; Khummongkol and Yokota, 2016). This is one kind of cross-media operation via intermediate $\boldsymbol{L}_{md}$ representation where sensory-motory information and conceptual information are computable uniformly. At his best knowledge, there is no other theory or system that can perform cross-media operations in such a seamless way as ours.

# Conclusions

This book presented an original semantic theory named mental image directed semantic theory and its application to a natural language understanding methodology called mental-image based understanding, intending for robots to understand texts in the same way that people do. This is quite distinguished from conventional theories and shows a potential good enough to be a very powerful means for realizing robotic awareness in computer and its understanding. To our best knowledge, there is no research similar to ours, namely, natural language understanding based on the model of mental image processing. Therefore, we cannot present any quantitative comparison with others. However, judging from the evaluation of conversation management system based on our psychological experiment, mental image directed semantic theory could provide the mental-image based understanding system, namely, conversation management system with an effective methodology to return the correct and satisfactory human-like answers in question-answering.

The system was designed to disambiguate an input sentence for its most plausible *semantic* interpretation by semantic computation (i.e., inference) in $L_{md}$. Disambiguation is the most serious problem for any natural language processing system. Most current approaches to it are based on the statistics about certain corpora of texts, however, they are what lead to the most plausible *syntactic* interpretation but not to the most plausible *semantic* interpretation grounded in the concerned world that is most essential to work robots appropriately by words.

Anna, an imaginary robot still under development, has been implemented in conversation management system as an artificial intelligence managing conversation in natural language and playing in animation. Her reasoning is based on her conception in semantics for humans and, therefore, she must interpret her solution in semantics for robots when she, as a real robot, acts it out in order to control her actuators/sensors properly. The semantic computation in $L_{md}$ performed by her is based on simple and general rules about atomic loci, hence, conversation management system works feasibly in Python except for computational cost in the Animation Generator.

Considering that human selective attention mechanism yielding semantic articulation should greatly concern the differences between deep neural networks and human vision, this volume presented a model of human active perception based on mental image directed semantic theory and conversation management system as an example of systematic control of robotic awareness based on the knowledge acquired through the model. The psychological experiment showed that conversation management system has the potential for people to control robotic awareness systematically with natural language. From the viewpoint of machine learning, the author believes that this research can also give a good suggestion about how robots should acquire knowledge so as to be useful for higher level cognition, such as abstract reasoning based on both linguistic and non-linguistic knowledge.

Another aim of our research is to target problem-finding/solving by human-robot cooperation, especially through communication in human language (i.e., natural language), where the core technologies are natural language understanding and knowledge management based on the formal language $\boldsymbol{L}_{md}$. Conventional methodologies for problem-solving inevitably employ simple state-space models or so because of lacking knowledge representation languages, such as $\boldsymbol{L}_{md}$, capable of formulating complex and dynamic problems in seamless connection with natural language understanding.

As for the coverage of word concepts, the author has analyzed a considerable number of spatial terms over various kinds of English words, such as prepositions, verbs, adverbs, etc., categorized as Dimensions, Form and Motion in the class SPACE of the Roget's thesaurus, and found that almost all the concepts of 4D events can be defined in exclusive use of 5 kinds of attributes for the focus of attention of the observer, namely, Physical location ($A_{12}$), Direction ($A_{13}$), Trajectory ($A_{15}$), Mileage ($A_{17}$) and Topology ($A_{44}$). This implies that spatiotemporal information systems with natural language interfaces are very feasible in terms of the size of knowledge to be installed.

## Acknowledgements

This work was partially funded by the Grants from Ministry of Education, Culture, Sports, Science and Technology, Japanese Government, numbered 62580024, 62210013, 62210010, 63633519, 01633014, 03245221, 05241105, 07207124, 08207118, 08680419, 09204110, 10111106, 14580436, 17500132, 23500195, and 19K12109.

# References


Alissandrakis A., Nehaniv C.L. and Dautenhahn K. (2003). Imitating with ALICE: Learning to imitate corresponding actions across dissimilar embodiments. IEEE Transactions on Systems, Man and Cybernetics, Part A: Systems and Humans, 32-4: 482–496.

Allen J.F. (1984). Towards a general theory of action and time. Artificial Intelligence, 23-2: 123–154.

Bahdanau D., Cho K. and Bengio Y. (2015). Neural machine translation by jointly learning to align and translate. Proc. of ICLR 2015.

Baldwin T., Dras M., Hockenmaier J., King T.H. and van Noord G. (2007). The impact of deep linguistic processing on parsing technology. In Proc. of the 10th International Workshop on Parsing Technologies (IWPT-2007), pp. 36–8, Prague, Czech Republic.

Billard A. (2000). Learning motor skills by imitation: biologically inspired robotic model. Cybernetics and Systems, 32: 155–193.

Brachman R. (1985). Introduction. pp. XVI–XVII. *In*: Brachman R. and Levesque H.J. (eds.). Readings in Knowledge Representation. Morgan Kaufmann.

Brachman R. and Schmolze J.G. (1985). An Overview of the KL-ONE Knowledge Representation System. Cognitive Science, 9: 171–216.

Chandioux J. (1976). METEO. FBIS seminar on MT, American Journal of Computational Linguistics, 46.

Chomsky N. (1956). Three models for the description of language. IRE Transactions on Information Theory, 2-3: 113–124.

Cohen R. (1984). A computational theory of the function of clue words in argument understanding. In Proceedings of the 10th international conference on computational linguistics, pp. 251–258.

Copeland, B.J. (2018). Artificial intelligence-Strong AI, applied AI, and cognitive simulation. Encyclopædia Britannica.

Coradeschi S. and Saffiotti A. (2003). An introduction to the anchoring problem. Robotics and Autonomous Systems, 43: 85–96.

Coventry K.R., Prat-Sala M. and Richards L.V. (2001). The interplay between geometry and function in the comprehension of 'over', 'under', 'above' and 'below'. Journal of Memory and Language, 44: 376–398.

Dagan I., Glickman O. and Magnini B. (2006). The PASCAL Recognizing Textual Entailment Challenge. *In*: Machine Learning Challenges, Springer Verlag, LNAI 3944.

Davidson D. (1967). The Logical Form of Action Sentences. pp. 81–95, University of Pittsburgh Press, Pittsburgh, Pennsylvania.

Davis R., Shrobe H. and Szolovits P. (1993). What Is a Knowledge Representation? AI Magazine. 14-1: 17–33.

Demiris Y. and Khadhouri B. (2006). Hierarchical attentive multiple models for execution and recognition of actions. Robotics and Autonomous Systems, 54: 361–369.

Drumwright E., Ng-Thow-Hing V. and Mataric M.J. (2006). Toward a vocabulary of primitive task programs for humanoid robots. Proc. of International Conference on Development and Learning (ICDL06), Bloomington IN, May 2006.

Egenhofer M. (1991). Point-set topological spatial relations. Geographical Information Systems, 5-2: 161–174.

Ferrucci D. and Lally A. (2004). UIMA: An architectural approach to unstructured information processing in the corporate research environment. Journal of Natural Language Engineering.

Ferrucci D., Brown E., Chu-Carroll J., Fan J., Gondek D., Kalyanpur A.A., Lally A., Murdock J.W., Nyberg E., Prager J., Schlaefer N. and Welty C. (2010). Building Watson: An Overview of the DeepQA Project. AAAI AI Magazine.

Fillmore, C.J. (1968). The case for case. pp. 1–88. *In*: Bach and Harms (ed.). Universals in Linguistic Theory. New York: Holt, Rinehart, and Winston.

Frege, G. (1879). On sense and reference. pp. 56–78. *In*: Translations from the Philosophical Writings of Gottlob Frege (2nd ed.). Peter Geach P. and Black M. (eds.). Oxford: Basil Blackwell, 1960.

Grice H.P. (1975). Logic and conversation. pp. 41–58. *In*: Cole P. and Morgan J.L. (eds.). Speech Acts. New York: Academic Press.

Guha R.V. and Lenat D.B. (1990). CYC: A Mid-Term Report. AI Magazine, 11-3: 32–59.

Han A.L., Wong D.F. and Chao L.S. (2012). LEPOR: A robust evaluation metric for machine translation with augmented factors. pp. 441–450. *In*: Proceedings of the 24th International Conference on Computational Linguistics (COLING 2012), Mumbai, India.

Harnad S. (1990). The Symbol Grounding Problem. Physica D 42, pp. 335–346.

Harding J. (2002). Geo-ontology Concepts and Issues. Report of a workshop on Geo-ontology, Ilkley UK.

Hays D.G. (1967). Introduction to Computational Linguistics. American Elsevier, New York.

Heylighen F. (1988). Formulating the Problem of Problem-Formulation. pp. 949–957. *In*: Trappl R. (ed.). Cybernetics and Systems '88, Kluwer Academic Publishers, Dordrecht.

Hobbs J.R. (1985). Ontological promiscuity. *In*: Proc. of ACL, pp. 60–69.

Katz G.E., Huang D., Gentili R. and Reggia J. (2016). Imitation learning as cause-effect reasoning. Proceedings of the 9th Conference on Artificial General Intelligence (AGI-16), New York, Jul. 2016.

Katz J.J. and Fodor J.A. (1963). The structure of a semantic theory. Language, 39-2, Apr–Jun, pp. 170–210.

Kawaguchi E., Yokota M., Endo T. and Tamati T. (1979). An understanding system of natural language and pictorial pattern in the world of weather report. Proceedings of the 6th International Joint Conference on Artificial Intelligence, 1, Tokyo Japan, Aug. 1979.

Khummongkol R. and Yokota M. (2014). Simulation of human awareness control in spatiotemporal language understanding as mental image processing. Proc. of The IEEE Symposium Series on Computational Intelligence, Electronic ISBN: 978-1-4799-4476-7.

Khummongkol R. and Yokota M. (2016). Computer simulation of human-robot interaction through natural language. Artificial Life and Robotics. doi:10.1007/s10015-016-0306-5, pp. 1-10, Springer-Verlag.

Kishimoto Z. (2016). What's Wrong Current Approaches to AI? URL=http://tek-tips.nethawk.net/what%E2%80%99s-wrong-current-approaches-to-ai/.

Koch C. (2004). The Quest for Consciousness: A Neurobiological Approach. Englewood, CO: Roberts & Company.

Koehn P. (2009). Statistical Machine Translation. Cambridge University Press.

Langacker R.W. (2005). Dynamicity, fictivity, and scanning: The imaginative basis of logic and linguistic meaning. pp. 164–197. *In*: Pecher D. and Rolf Zwaan A. (eds.). Grounding Cognition: The Role of Perception and Action in Memory Grounding Cognition: The Role of Perception and Action in Memory. Language, and Thinking.

Leisi E. (1961). Der Wortinhalt: Seine Struktur im Deutschen und Englischen. Quelle & Meyer, Heidelberg.

Lenat D.B. and Guha R.V. (1990). Building Large Knowledge-Based Systems: Representation and Inference in the Cyc Project. Addison-Wesley, MA.

Levesque H. and Lakemeyer G. (2008). Cognitive robotics. *In*: Van Harmelen F. and Porter B. (eds). Handbook of Knowledge Representation. Foundations of Artificial Intelligence, 3: 869–886, Elsevier, Amsterdam.

Levesque H.J. (2011). The Winograd schema challenge. AAAI Spring Symposium: Logical Formalizations of Commonsense Reasoning, Palo Alto CA, March 2011.

Levesque H.J. (2014). On our best behavior. Artificial Intelligence, 212: 27–35.

Logan G.D. and Sadler D.D. (1996). A computational analysis of the apprehension of spatial relations. pp. 493–529. *In*: Bloom P., Peterson M.A., Nadel L. and Garrett M. (eds.). Language and Space. Cambridge, MA: MIT Press.

Manning C.D. and Schütze H. (1999). Foundations of Statistical Natural Language Processing. MIT Press.

McDermott D.V. (1982). A temporal logic for reasoning about processes and plans. Cognitive Science, 6: 101–155.

Miller G.A. and Johnson-Laird P.N. (1976). Language and Perception. Harvard University Press.

Miller G.A., Beckwith R., Fellbaum C.D., Gross D. and Miller K. (1990). WordNet: An online lexical database. Int. J. Lexicograph. 3, 4: 235–244.

Minsky M. (1986). The society of mind. Simon and Schuster, New York.

Morris C. (1938). Foundations of the Theory of Signs. *In*: Neurath O. (ed.). International Encyclopedia of Unified Science, 1-2.

Nagao M. (1984). A framework of a mechanical translation between Japanese and English by analogy principle. *In*: Elithorn A. and Banerji R. (eds.). Artificial and Human Intelligence. Elsevier Science Publishers.

Nakanishi J., Morimoto J., Endo G., Cheng G., Schaal S. and Kawato M. (2004). Learning from demonstration and adaptation of biped locomotion. Robotics and Autonomous Systems, 47(2-3): 79–81.

Navigli R. (2009). Word Sense Disambiguation. A Survey, ACM Computing Surveys, 41-2: 1-69.

Nguyen A., Yosinski J. and Clune J. (2015). Deep Neural Networks are Easily Fooled: High Confidence Predictions for Unrecognizable Images. CVPR'15, 2015.

Noton D. (1970). A theory of visual pattern perception. IEEE Transaction on Systems Science and Cybernetics, Vol. SSC-6, No. 4, pp. 349–357.

Nyberg E. and Mitamura T. (1992). The KANT System: Fast, Accurate, High-Quality Translation in Practical Domains, Proceedings of COLING'92, Nantes, France, July 1992.

Paivio A. (1971). Imagery and verbal processes. New York: Holt, Rinehart, and Winston.

Prinz W. (1984). Modes of linkage between perception and action. *In*: Prinz W. and Sanders A.F. (eds.). Cognition and Motor Processes. Springer, Berlin, Heidelberg.

Prinz W. (1990). A common coding approach to perception and action. *In*: Neumann O. and Prinz W. (eds.). Relationships Between Perception and Action. Springer, Berlin, Heidelberg.

Roget P. (1975). Thesaurus of English Words and Phrases. J.M. Dent and Sons Ltd., London.

Rouchota V. (1996). Discourse Connectives: What do They Link? UCL Working Papers in Linguistics, 8: 199–214.

Roy D. (2005). Semiotic Schemas: A framework for grounding language in action and perception. Artificial Intelligence, 167: 170–205.

Russell S.J. and Norvig P. (2010). Artificial intelligence: A modern approach (3rd ed.). Upper Saddle River: Prentice Hall.

Rysová M. and Rysová K. (2018). Primary and secondary discourse connectives: Constraints and preferences. Journal of Pragmatics, 130: 16–32.

Schank R.C. (1969). A conceptual dependency parser for natural language. Proc. of the 1969 conference on Computational linguistics, Sång-Säby, Sweden, pp. 1–3.

Schank R.C. and Abelson R.P. (1977). Scripts, Plans, Goals and Understanding. Hillsdale, NJ: Lawrence Erlbaum.

Schiffrin D. (1987). Discourse markers. Cambridge University press.

Schlaefer N. and Welty C. (2010). Building Watson: An Overview of the DeepQA Project. AAAI AI Magazine (Fall 2010).

Schubert L. (2015a). Semantic Representation. 29th AAAI Conference, Austin TX.

Schubert L. (2015b). Computational Linguistics. The Stanford Encyclopedia of Philosophy (Spring 2015 Edition), *In*: Zalta E. N. (ed.). URL = https://plato.stanford.edu/archives/spr2015/entries/computational-linguistics/.

Sekine S., Uchimoto K. and Isahara H. (2000). Backward beam search algorithm for dependency analysis of Japnese. *In*: Proceedings of Computational Linguistics (COLING), pp. 754–760.

Shariff A.R., Egenhofer M. and Mark D. (1998). Natural-language spatial relations between linear and areal objects: The topology and metric of english-language terms. International Journal of Geographical Information Science, 12-3: 215–246.

Shepard, R.N. and Metzler J. (1971). Mental rotation of three-dimensional objects. Science, 171(3972): 701–703.

Shiraishi M., Capi G. and Yokota M. (2006). Human-robot communication based on a mind model. Artificial Life and Robotics, 10(2): 136–140.

Sowa J.F. (2000). Knowledge Representation: Logical, Philosophical, and Computational Foundations. Brooks Cole Publishing Co., Pacific Grove, CA.

Swan M. (2005). Practical english usage. Oxford: Oxford University Press.

Taniguchi R., Yokota M., Kawaguchi E. and Tamati T. (1984). Knowledge-based picture understanding of weather charts. Pattern Recognition, 17-1: 109–123, Elsevier.

Tauchi E., Sugita K., Capi G. and Yokota M. (2006). Towards artificial communication partners with kansei. Proceedings of IEEE workshop on Network-based Virtual Reality and Tele-existence (INVITE06), Wien Austria, May 2006.

Thagard P. (2014). Cognitive science, the stanford encyclopedia of philosophy (Fall 2014 Edition). *In*: Zalta E.N. (ed.). URL = https://plato.stanford.edu/archives/fall2014/entries/cognitive-science/.

Thomas N.J.T. (2018). Mental imagery. *In*: Zalta E.N. (ed.). The Stanford Encyclopedia of Philosophy (Spring 2018 Edition). URL =https://plato.stanford.edu/archives/spr2018/entries/mental-imagery/.

Thompson C. (2010). What Is I.B.M.'s Watson? The New York Times Magazine (June 16, 2010).

Tongchim S., Altmeyer R., Sornlertlamvanich V. and Isahara H. (2008). A dependency parser for thai. Proceedings of the International Conference on Language Resources and Evaluation, LREC 2008, 26 May–1 June 2008, Marrakech, Morocco.

Tsarkov D. and Horrocks I. (2006). FaCT++ Description Logic Reasoner: System Description. Automated Reasoning, Lecture Notes in Computer Science, 4130: 292–297.

Turing A.M. (1950). Computing Machinery and Intelligence. Mind 49: 433–460.

Uexküll J. and Kriszat G. (1934). Streifzüge durch die Umwelten von Tieren und Menschen: Ein Bilderbuch unsichtbarer Welten, Julius Springer in Berlin.

Wächtera M., Ovchinnikovaa E., Wittenbecka V., Kaisera P., Szedmakd S., Mustafac W., Kraftc D., Krügerc N., Piaterb J. and Asfoura T. (2018). Integrating multi-purpose natural language understanding, robot's memory, and symbolic planning for task execution in humanoid robots. Robotics and Autonomous Systems, 99: 148–165.

Wilks Y. (1972). Grammar, meaning and the machine analysis of language. London and Boston, Routledge.

Wilson R.A. and Foglia L. (2017). Embodied cognition. The stanford encyclopedia of philosophy (Spring 2017 Edition). *In*: Zalta E.N. (ed.). URL = https://plato.stanford.edu/archives/spr2017/entries/embodied-cognition/.

Winograd T. (1972). Understanding Natural Language. Academic Press.

Winograd T. (2006). Shifting viewpoints: Artificial intelligence and human–computer interaction. Artificial Intelligence, 170: 1256–1258.

Wolfe J.M. (1994). Visual search in continuous, naturalistic stimuli. Vision Research, 34: 1187–1195.

Woods W. (1970). Transition network grammars for natural language analysis. Communications of the ACM, 13-10: 591–606.

Woods W. (1975). What's in a link: Foundations for semantic networks. *In*: Bobrow D. and Collins A. (eds.). Representation and Understanding: Studies in Cognitive Science, pp. 35–82, New York: Academic Press.

Yatsko V.A., Starikov M.S. and Butakov A.V. (2010). Automatic genre recognition and adaptive text summarization. In: Automatic Documentation and Mathematical Linguistics, 44-3: 111–120.

Yokota M., Taniguchi R. and Kawaguchi E. (1984a). Language-picture question-answering through common semantic representation and its application to the world of weather report. pp. 203–253. *In*: Bolc L. (eds.). Natural Language Communication with Pictorial Information Systems. Symbolic Computation (Artificial Intelligence), Springer, Berlin, Heidelberg.

Yokota M., Yoshitake H. and Tamati T. (1984b). Japanese-English translation of weather reports by ISOBAR. Trans. IECE Japan E67/6, pp. 315–322.

Yokota M. (1999). Semantic analysis and description of conjunctive words for discourse synthesis. (in Japanese) Technical report of IEICE. Thought and language, 99-487: 11–16.

Yokota M. (2005). An approach to natural language understanding based on mental image model. pp. 22–31. *In*: Sharp B. (ed.). Natural Language Understanding and Cognitive Science, INSTICC PRESS.

Yokota M. and Capi G. (2005a). Cross-media operations between text and picture based on mental image directed semantic theory. WSEAS Transaction on Information Science and Applications, 10(2): 1541–1550.

Yokota M. and Capi G. (2005b). Integrated multimedia understanding for ubiquitous intelligence based on mental image directed semantic theory. Proceedings of IFIP EUC`2005 symposium (LNCS 3823), pp. 538–546, Nagasaki Japan, Dec., 2005.

Yokota M. (2006). Towards a Universal Language for Distributed Intelligent Robot Networking. Proceedings of International Conference on Systems, Man and Cybernetics (SMC06), Taipei Taiwan, Oct. 2006.

Yokota M. (2007). A theoretical consideration on robotic imitation of human action according to demonstration plus suggestion. Proceedings of the 4th International Symposium on Imitation in Animals and Artifacts Newcastle upon Tyne, April 2007.

Yokota M. (2008). A general theory of tempo-logical connectives and its application to spatiotemporal reasoning. Proc. of Joint 4th International Conference on Soft Computing and Intelligent Systems and Intelligent Systems and 9th International Symposium on Advanced Intelligent Systems, Nagoya Japan, Sep. 2008.

Yokota M., Shiraishi M., Sugita K. and Oka T. (2008). Toward integrated multimedia understanding for intuitive human-system interaction, Artificial Life and Robotics, 12(1-2): 188–193.

Yokota M. (2009). Systematic formulation and computation of subjective spatiotemporal knowledge based on mental image directed semantic theory: Toward a formal system for natural intelligence. Proc. of NLPCS2009, pp. 133–143, Milan Italy, May 2009.

Yokota M. (2010). Towards awareness computing under control by world knowledge grounded in sensory data. Proc. of IEEE SMC 2010, Istanbul Turkey, Oct. 2010.

Yokota M. (2012). Aware computing in spatial language understanding guided by cognitively inspired knowledge representation. Applied Computational Intelligence and Soft Computing. vol. 2012 Article ID 184103, 10 pages, doi:10.1155/2012/184103.

Yokota M. and Khummongkol R. (2014). Representation and computation of human intuitive spatiotemporal concepts as mental Imagery. Proc. of The IEEE 6th International Conference on Awareness Science and Technology Paris France, Oct. 2014.

Zeeman E.C. (1962). The topology of the brain and visual perception. pp. 240–248. *In*: The Topology of 3-manifolds, Prentice Hall, Englewood, NJ.

# Appendix

**Table A-1.** Look-up table for abbreviated expressions in alphabetical order and chapters for reference.

| Abbreviation | Full expression | Chapter |
|---|---|---|
| CAND | Consecutive AND | 5 |
| $D_{EE}$ | Definition of empty event | 6 |
| DG | Dependency grammar | 3 |
| $D_{RO}$ | Definition of reversal operation of spatial change event | 7 |
| DS | Dependency structure | 3 |
| $D_{TI}$ | Definition of time-interval | 7 |
| $D_{TL}$ | Definition of tempo-logical connectives | 6 |
| Em | Emotion processing agent | 4 |
| $I_{CS}$ | Commutativity law of SAND | 7 |
| $I_{EC}$ | Elimination law of CAND | 7 |
| $I_{ES}$ | Elimination law of SAND | 7 |
| In | Integration processing agent | 4 |
| $I_{SL}$ | Substitution law of locus formulas | 7 |
| ISR | Imaginary space region | 7 |
| Kn | Knowledge processing agent | 4 |
| KRL | Knowledge representation language | 1 |
| $L_{md}$ | Mental image description language | 1 |
| MIDST | Mental image directed semantic theory | 1 |
| $P_{A1}$ | Postulate of arbitrariness in locus articulation-type1 | 7 |
| $P_{A2}$ | Postulate of arbitrariness in locus articulation-type2 | 7 |
| $P_{at}$ | Postulate of preposition at | 8 |

*Table A-1. contd. ...*

... *Table A-1. contd.*

| Abbreviation | Full expression | Chapter |
|---|---|---|
| $P_{CT}$ | Postulate of compound time-interval | 7 |
| $P_{CV}$ | Postulate of conservation of values | 7 |
| $P_{IS}$ | Postulate of arbitrary production of imaginary space region | 7 |
| PL | Predicate logic | 7 |
| $P_{MS}$ | Postulate of matter as standard | 7 |
| $P_{MV}$ | Postulate of matter as value | 7 |
| $P_{NT}$ | Postulate of negation of time-interval | 7 |
| $P_{PM}$ | Postulate of partiality of matter | 7 |
| $P_{RS}$ | Postulate of reversibility of spatial change event | 7 |
| $P_{SC}$ | Postulate of shortcut of causal chain | 7 |
| PSG | Phrase structure grammar | 3 |
| $P_{V1}$ | Postulate of identity of assigned values-type1 | 7 |
| $P_{V2}$ | Postulate of identity of assigned values-type2 | 7 |
| Re | Response processing agent | 4 |
| SAND | Simultaneous AND | 5 |
| St | Stimulus processing agent | 4 |
| $T_{CE}$ | Theorem of coexistence of empty event | 7 |
| $T_{TC}$ | Tempo-logical contrapositive | 7 |
| $T_{TP}$ | Theorem of absoluteness of time passing | 7 |

# Index

Q



R



S



T



U



W

# Color Section

## Chapter 3: Fig. 3.3, p. 31

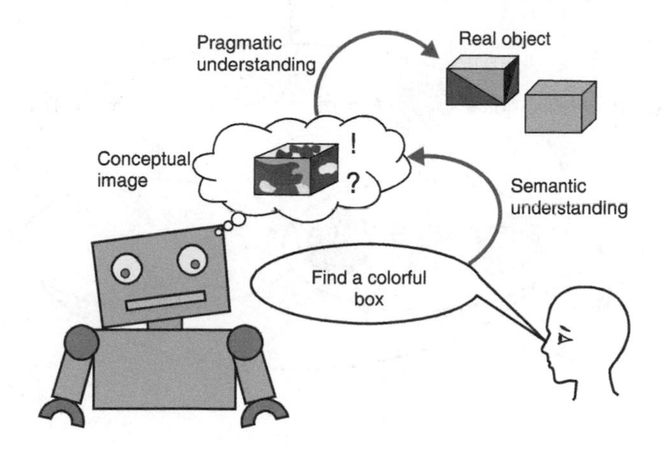


## Chapter 4: Fig. 4.9, p. 57

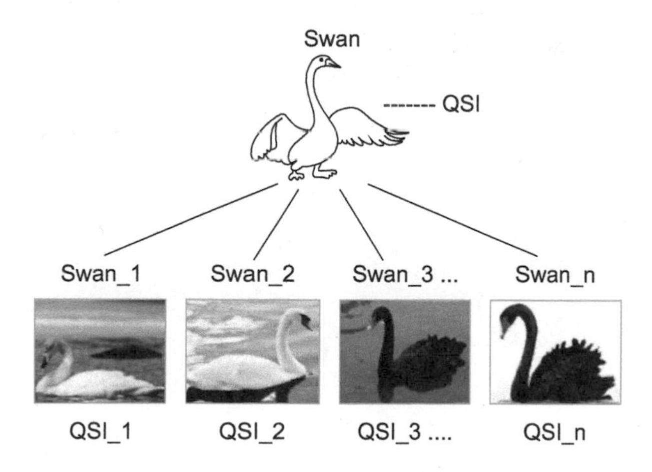

**Chapter 4: Fig. 4.11, p. 58**

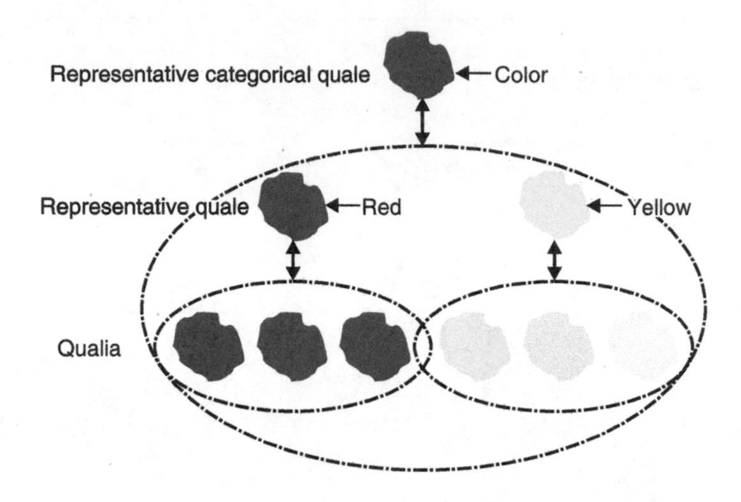


**Chapter 5: Fig. 5.3, p. 68**

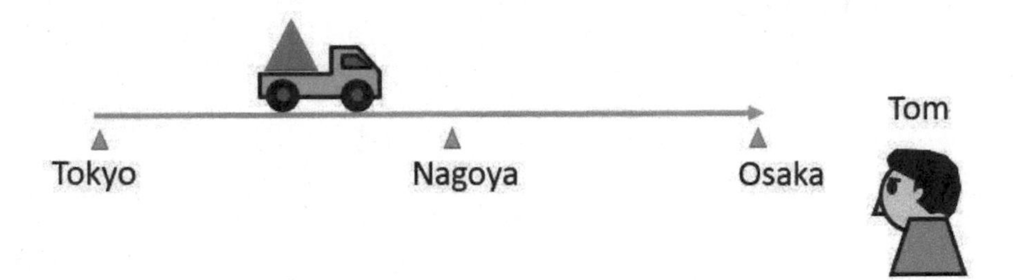